Water Privatisation

Water Privatisation

Trans-national corporations and the re-regulation of the water industry

Matthias Finger and Jeremy Allouche

SPON PRESS
Taylor & Francis Group

London and New York

First published 2002
by Spon Press
11 New Fetter Lane, London EC4P 4EE

Simultaneously published in the USA and Canada
by Spon Press
29 West 35th Street, New York, NY 10001

Spon Press is an imprint of the Taylor & Francis Group

Typeset in Times by
BOOK NOW Ltd
Printed and bound in Great Britain by
University Press, Cambridge

British Library Cataloguing in Publication Data
A catalogue record for this book is available
from the British Library

Library of Congress Cataloging in Publication Data
Finger, Matthias
Water privatisation : trans-national corporations and the re-regulation of the water industry / Matthias Finger and Jeremy Allouche.
p. cm.
1. Water utilities—Deregulation. I. Allouche, Jeremy. II. Title.

HD4456.F564 2001
363.6′1–dc21 2001023735

ISBN 0-415-23208-2

Contents

Spon's Environmental Science and Engineering Series

This new series covers a wide range of water, waste and contaminated land issues in the context of current best practice, perception and legislation.

Series coverage is broad. On the water and wastewater side it includes resource management, treatment, distribution and collection, monitoring and regulation. In relation to waste management it extends across the waste hierarchy, covering a range of technical and regulatory issues in areas such as waste minimisation, separation and sorting, recycling and disposal, and the determination of appropriate waste strategy.

The series is targeted at engineers and scientists in the process, waste, and environmental sectors. Titles will also be of interest to economists, lawyers, legislators, regulators, and advanced students.

Series Editor: Jeremy Joseph
11 Mallory Avenue, Caversham, Reading, Berkshire RG4 6QN
email: jbjoseph@cwcom.net

Topics under consideration for the series include:

- Potable/usable water
- Waste and polluted waters and sludges
- Water control and management issues
- Contaminated land
- Non-disposal strategies for waste
- Waste disposal routes
- Wastes from agriculture

The series editor will be pleased to hear from potential authors interested in writing on any topics relevant to the series including, but not limited to, the issues cited above. Please contact him at the above address with an indication of the scope of any proposed volume together with details regarding its intended readership.

Contact may also be made with: Richard Whitby
Commissioning editor, Spon Press, 11 New Fetter Lane,
London EC4P 4EE, UK
email: richard.whitby@tandf.co.uk

All volumes are published by Spon Press, part of the Taylor and Francis Group, and are sold through our worldwide distribution networks.

Visit us on the web at: www.sponpress.com

Acknowledgements

The author and publishers wish to thank the following who have kindly given permission for the use of copyright material.

Association pour une taxation des Transactions financières pour l'Aide aux Citoyens.

Confederation for Unity, Recognition and Advancement of Government Employees.

© The Economist Newspaper Limited, London (27/1/01).

Dunning, John, *Multinational Enterprises and the Global Economy*, Addison-Wesley Publications, © Addison Wesley Publishers 1993.

Food and Agriculture Organization of the United Nations (*The State of Food and Agriculture*, 1993).

Fortune 500, © 2000 Time Inc, all rights reserved.

Global Water Partnership (*Technical Advisory Committee Background Paper* 1 and 2).

International Labour Organisation (Labour and social dimensions of privatization and restructuring: public utilities (water, gas, electricity): 122, © ILO, 1998).

OECD Environmental Data: Compendium 1999. © OECD, 1999.

Office Internationale de l'Eau.

Public Services Privatisation Research Unit.

Reason Public Policy Institute #151 (October 1992) and Reason Public Policy Institute Study #162 (July 1993) reprinted, with permission. © 2001 by the Reason Foundation, 3415 S. Sepulveda Blvd, Suite 400, Los Angeles, CA 90034 <www.rppi.org>.

United Nations © Copyright United Nations 1997.

World Bank (PPI Project Database <www.worldbank.org/html/fpd/privatesector/ppidb.htm>).

Tony Allan and Richard Franceys have kindly provided advice or written material.

Abbreviations and acronyms

ACC	Administrative Committee on Coordination (UN)
ADB	Asian Development Bank
AGM	Agbar Global Market
ANDA	*Administracion Nacional de Acueductos (El Salvador)*
AP	Associated Programme (GWP)
ARA	Autonomous River Authority (France)
ATTAC	Association for the Taxation of Financial Transactions for the Aid of Citizens
ATV	*Abwassertechnische Vereinigung*
BAT	best available technology
BMZ	*Bundesministerium für wirtschaftliche Zusammenarbeit und Entwicklung*
BOOT	Build–Own–Operate–Transfer
BOT	Build–Operate–Transfer
CACB	Cooperative and Agricultural Credit Bank
CEE	Central and Eastern Europe
CGEA	*Compagnie Générale de Chauffe*
CIDA	Canadian International Development Agency
COURAGE	Confederation for Unity, Recognition and Advancement of Government Employees (Philippines)
CSC	Customer Service Committee
CVP	Central Valley Project (California)
DANIDA	Danish Development Agency
DBO	Design–Build–Operate
DFID	British Development Agency
DIREN	*Directions Régionales de l'Environnement* (France)
DVGW	*Deutscher Verein des Gas und Wasserfaches*
DVWK	*Deutscher Verband für Wasserwirtschaft und Kulturbau*
DWI	Drinking-Water Inspectorate (UK)
EBRD	European Bank for Reconstruction and Development
ECJ	European Court of Justice

ECLAC	Economic Commission for Latin America and the Caribbean (UN)
ECOSOC	Economic and Social Council (UN)
EECI	*Energie Electrique de Côte d'Ivoire*
EPA	Environmental Protection Agency (USA)
ESD	Environmentally Sustainable Development
EU	European Union
FAO	Food and Agriculture Organisation (UN)
FDI	foreign direct investment
GEF	Global Environmental Facility (World Bank)
GNP	gross national product
GWP	Global Water Partnership
Habitat	United Nations Center for Human Settlement
HCH	*Haut Commissariat à l'Hydraulique*
HWRP	Hydrology and Water Resources Programme (WHO)
IADB	Inter-American Development Bank
IAEA	International Atomic Energy Agency (UN)
IAHR	International Association for Hydraulic Research
IAHS	International Asociation of Hydrological Sciences
IAWQ	International Association on Water Quality
ICID	International Commission on Irrigation and Drainage
IFC	International Finance Corporation
IFI	International Financial Institution
IFPRI	International Food Policy Research Institute (USA)
IHE	International Institute for Infrastructural, Hydraulic and Environmental Engineering (Delft, Netherlands)
IHP	International Hydrological Programme (UNESCO)
IIASA	International Institute for Applied Systems Analysis
ILO	International Labor Organisation
INBO	International Network of Basin Organisation (France)
INSTRAW	International Research and Training Center for the Advancement of Women
IOC	Intergovernmental Oceanographic Commission
ISO	International Organisation for Standardisation
IUCN	World Conservation Union (Switzerland)
IWA	International Water Association (UK)
IWRA	International Water Resources Association (USA)
JMP	Joint Monitoring Programme for Water Supply and Sanitation (UNESCO/WHO)
LWUA	Local Water Utilities Administration (Philippines)
M&A	merger and acquisition
MISE	*Mission InterServices de l'Eau* (France)
MNC	multinational corporation
MNE	multinational enterprise

MWC	Manila Water Company, Inc. (Philippines)
MWS	Manylad Water Services, Inc. (Philippines)
MWSS	Metro Manila's Water Sewerage System (Philippines)
NAFTA	North American Free Trade Agreement
NCC	National Customer Council (OFWAT, UK)
NGO	non-governmental organisation
NPM	new public management
NRA	National Rivers Authority
NTB	non-tariff barrier
NWRB	National Water Resources Board (Philippines)
NWW	North West Water (UK)
ODA	Overseas Development Agency (UK)
ODI	Overseas Development Institute (UK)
OECD	Organisation for Economic Cooperation and Development
OED	Operations Evaluations Department (World Bank)
OFWAT	Office of Water Services (UK)
OMM	*Organisation Météorologique Mondiale*
OTV	*Omnium de Traitement et de Valorisation*
PPP	'polluter-pays' principle
PPP	Public–Private Partnership
PSP	Private Sector Participation
PSPRU	Public Services Privatisation Research Unit
PSTNC	Public Sector Transnational Corporations
RAD	*Régie Autonome de Distribution*
R&D	research and development
RO	Regulatory Office (Philippines)
SAP	Structural Adjustment Policy (World Bank)
SEG	*Société des Eaux de Grenoble*
SGE	*Société Générale d'Entreprise*
SIDA	Swedish Development Agency
SIWI	Stockholm International Water Institute
SODECI	*Société de distribution d'eau Côte d'Ivoire*
SOE	state-owned enterprise
SWP	State Water Projects (California, USA)
TNC	transnational corporation
TNE	transnational enterprise
UADE	*Union Africaine pour le Développement de l'Eau*
UN	United Nations
UNCHS	United Nations Center for Human Settlements
UNCNR	United Nations Committee on Natural Resources
UNCSD	United Nations Commission on Sustainable Development
UNCTAD	United Nations Conference for Trade and Development
UNCTC	United Nations Center on Transnational Corporations
UNDP	United Nations Development Programme

UNEP	United Nations Environment Programme
UNESCO	United Nations Educational, Scientific and Cultural Organisation
UNICEF	United Nations Children's Fund
USSR	Union of Soviet Socialist Republics
VAT	value added tax
WASAMS	Water and Sanitation Monitoring System
WCC	World Water Council
WCD	World Commission on Dams
WEDC	Water, Engineering and Development Center
WHO	World Health Organisation
WMO	World Meteorological Organisation
WRAC	Water Resources Advisory Committee
WRI	World Resources Institute (USA)
WSP	World Bank Sanitation Programme
WTO	World Trade Organisation

Introduction

This is a book about the evolution of water resources management in the age of globalisation. In it we show how, over recent years, water has evolved from being a common good and a public service into a commodity that is increasingly being managed according to economic principles. This evolution, of course, goes hand in hand with a change from public water utilities to transnational corporations, the so-called TNCs. Moreover, all this must be seen against the background of economic globalisation, the changing role of the nation-state, and, above all, the pursuit of industrial development. In light of all these evolutions, this book, which analyses the different forces and actors in place is, we think, needed.

As a matter of fact, in the minds of many peoples water is (still) something very special, and as such reflects an ancestral view of our civilisations' relationships to one of the most vital resources: water, it is said, is an essential good, something that humans and cultures rely upon for their existence and development, not to mention the fact that water often has a religious meaning. In short, water, in overall perception, remains public and special, something that cannot be commodified, nor commercialised, nor traded. Yet, this is exactly what is about to happen, as a result of two separate trends, namely the simultaneous pursuit of industrial development and the transformation of the public sector.

Indeed, water is historically part and parcel of the industrial development paradigm. As such, water has been approached until recently mainly from a supply perspective along with the question of redistribution, i.e. when and where people wanted it for their use. This is also the so-called 'engineering' approach. However, one has to realise that water is an increasingly scarce resource in most parts of the world and the pressure on water resources will only grow with the pursuit of industrial development, increased urbanisation, concentration of population, and industrialised food production (irrigation). At the same time, one must keep in mind that although water needs to be put onto a more sustainable path, there are still more than 1.2 billion people lacking access to clean drinking water, and more than 2.9 billion people still lacking access to adequate sanitation facilities. In short, the pursuit of industrial development increasingly hits environmental and other limits of supply, thus gradually leading to a demand approach. This means that one also has to consider a better redistribution of water, so as to ensure universal access to this vital good.

Yet, in the context of a weakening State, the questions of water redistribution and environmental protection are increasingly approached from an economic perspective, rather than in purely political terms. Indeed, globalisation not only forces the nation-state and other political actors – traditionally in charge of water distribution – to position themselves in light of competing actors, but moreover gives rise to new and powerful actors in the water sector, such as global financial institutions and transnational corporations. Consequently, the discussions about water have gradually become a matter of anti- and pro-privatisation, i.e. a matter of institutional policy, rather than a matter of substance. As an illustration, the highly mediated World Water Forum in March 2000 turned into a confrontation between pro- and contra-water privatisers, a symptom of the pro- and anti-globalisation campaign.

From our perspective, however, it is not so much privatisation that should be questioned, as the coherence of the overall institutional arrangements of global resources water management. What is ultimately the best way to manage water in the age of globalisation, with many powerful corporations having the expertise to do so, while peoples and public entities increasingly lack the power to make their interests heard? How can decentralisation, user-participation, inter-regional co-operation, river basin management and other approaches enhancing local responsibilities be reconciled with growing financial needs, technological expertise and the growing power of transnational corporations in the sector? In short, how can an increasingly global concern with and interest for water be reconciled with the need and desire for control by local peoples? This is the intellectual and practical challenge we would like to address in this book.

We will do so from an organisational perspective, i.e. a perspective, which considers the different actors and interests involved and analyses their respective strategies. This kind of analysis is, we think, particularly pertinent in the water sector, as this sector is increasingly characterised, as a result of globalisation, by a multiplication of actors and organisational forms, such as networks, associations, organisations and private corporations. In this book, therefore, particular attention will be given to the transformation and emergence of actors at all levels, as well as to their (strategic) relationships and potential to solving the increasingly global 'water crisis'.

Consequently, we have decided to structure this book into seven separate chapters: in Chapter 1, we will highlight the overall trend of globalisation and analyse its impact on the nation-state, the rise of transnational corporations and, as a result, the changing approach to infrastructure management and development of water. In Chapter 2, we will identify the parallel emergence of environmental concerns leading most recently to considering water as an economic good. Chapters 3 and 4 will present the emergence of the two, in our opinion, most powerful actors in the water sector today and even more so in the future, namely the World Bank and some TNCs. We will analyse their behaviour and strategies, and speculate as to how their roles in this sector might evolve in the future. Chapters 5 and 6 will illustrate the evolution of water resources management in the case of selected examples in the South and the North. Indeed, it is in the South where these actors' strategies are currently being played out. Yet, we anticipate that the water sector will increasingly be shaped along the same lines – as well as by the same multinational corporations (MNCs) – also in the North, an evolution which can already be observed in France and the United Kingdom. Chapter 7, therefore, will outline alternatives and remedies to this, we think, fundamentally undesirable evolution, in terms of both citizens' participation and safeguards for public service provision. In short, this book wishes to highlight the current trends in global water management, as well as to outline realistic approaches as to how to get these trends back under public control.

Chapter 1

Globalisation and the state's changing role in infrastructure development

> In circumstances of accelerating globalisation, the nation-state has become 'too small for the big problems of life and too big for the small problems of life'.
>
> (Anthony Giddens)

The reader may be surprised to see this book, which focuses essentially on water management, start off by examining the implications of globalisation on the nation-state, the international state system and transnational actors. However, we think that such an analysis is necessary in order to better understand the current transformations in the water sector, given in particular the fact that water has now become a global issue. Indeed, much of the transformations the water sector and other network industries are currently undergoing have their origins in the very process of globalisation and related dynamics of deregulation and privatisation. In parallel, the state, which is being put under pressure by the same process of globalisation, is trying to readjust to this new situation by developing new ways and means to ensure both investments and regulation in the water and all other infrastructure sectors. Simultaneously, new international arrangements and corresponding global actors are emerging in this sector and elsewhere.

This chapter is structured in three sections: first we will examine the different dimensions of globalisation – especially the technological, financial and economic dimensions – and their implications for managing public affairs both nationally and globally. In a second section, we will then examine in particular the institutional dimensions of globalisation, and see how globalisation affects the international system and the state in general. Finally, we will analyse the consequences of the changing international system and the state on infrastructures more generally, especially on water infrastructures worldwide. As such, this introductory chapter seeks to lay out the overall conceptual framework within which, we think, the question of water and its management must ultimately be seen.

Globalisation

The purpose of this section is to foster a clearer understanding of the dynamics of globalisation, in particular its technological and economic dimensions. First, we want to conceptualise globalisation as a multi-faceted process, and not simply as an outcome, as is often done in the social science literature. Indeed, many authors confuse process and outcome, a confusion, which often leads them to judge whether such globalisation is desirable or not. Therefore, we will, in a second step, look more precisely into the technological as well as the economic dimensions of globalisation, as these dimensions are particularly relevant when it comes to infrastructure development. The next section will then focus in particular on the institutional dimensions of such globalisation.

Understanding globalisation

According to an Organisation for Economic Cooperation and Development (OECD) publication,[1] the term 'globalisation' was first used in 1985 by Theodore Levitt in respect of *The Globalisation of Markets* (Levitt, 1985). Levitt used the expression to characterise the vast changes that had taken place over the previous two decades in the international economy, i.e. the rapid and pervasive diffusion of production, consumption and investment of goods, services, capital and technology worldwide.

More precisely, globalisation was initially mainly used by economic historians in order to characterise the changing nature of the economy, which indeed remained the main meaning until very recently. However, increasingly this term is now being used in most of the social science literature in order to describe many different forms of change – i.e. from economic to political and even to social and cultural change. As such, this literature tends to become quite unspecific, as it also tends to confuse process and outcome, thus mixing descriptive with normative elements.

For example, globalisation, wrote Anthony Giddens in 1990, is 'a term which must have a key position in the lexicon of the social sciences' (Giddens, 1990: 52). More precisely, Giddens considers that

> In the modern era, the level of time–space distanciation is much higher than in any previous period, and the relations between local and distant social forms and events become correspondingly 'stretched'. Globalisation refers essentially to that stretching process, in so far as modes of

1 According to Ricardo Petrella, the term was first used by George Modelski in 1972, when referring to European-led expansion in order to gain control over other communities in the world and to integrate them into one single global trading system (Petrella, 1996: 63).

> connection between different social contexts or regions become networked across the earth's surface as a whole. Globalisation can thus be defined as the intensification of worldwide social relations which link distant localities in such a way that local happenings are shaped by events occurring many miles away and vice versa.
>
> (Giddens, 1990: 64)

McGrew similarly defines globalisation as 'the multiplicity of linkages and interconnections between the states and societies, which make up the present world system. It describes the process by which events, decisions, and activities in one part of the world come to have significant consequences for individuals and communities in quite distant parts of the globe' (McGrew and Lewis, 1992: 22). Both definitions pertain to the social and cultural dimensions of globalisation and stress the growing interconnectedness of the various social phenomena occurring around the globe, i.e. both a *process* and an *outcome*.

Similarly, in political science and international relations theory, globalisation usually points to the post-Westphalian era (e.g. Held, 1995) or, as Rosenau sees it, to a 'post-international' system (Rosenau, 1997). This means that, for political scientists and international relations specialists, globalisation defines both a process and a situation in which political relationships are less territorial-based, and nation-states become less important. Often, the term 'governance' is used here (Young and Hunold, 1999), in order to define a situation where nation-states are just one among several actors contributing to solving collective problems. We will return to this new institutional reality in the next section of this chapter.

However, before doing so, we must highlight the fact that globalisation has indeed become a fashion term with little specific content. Not surprisingly, Susan Strange considers 'globalisation' a term which can refer to anything from the internet to a hamburger. All too often, she says, it is a polite euphemism for the continuing Americanisation of consumer taste and cultural practices (Strange, 1996). Like Susan Strange, we think that one should analyse the process of globalisation in more detail, distinguishing, in particular, between the underlying forces of globalisation on the one hand and its – desirable or undesirable – outcomes (depending on one's point of view) on the other.

We do indeed see globalisation as the latest stage of a process of 'modernisation' of anthropological proportions. As such, this process has of course technological, economic, cultural, as well as institutional, social and political dimensions. We thus agree with Antony Giddens when he says that

> the declining grip of the West over the rest of the world is not the result of the diminishing impact of the institutions which first arose there but, on the contrary as a result of their global spread. The economic, political and

> military power which gave the West its primacy, . . . no longer so distinctly differentiates the Western countries from others elsewhere. We can interpret this phenomenon as one of globalisation.
>
> (Giddens, 1990: 52)

In short, globalisation must be seen, in our view, as a process by which Western technologies, production methods, institutions, consumption patterns and worldviews are increasingly spreading across the planet, to the point where they become more or less universal. The driving forces of this process are numerous, yet *technology*, *economics* and (political) *institutions* must, in our view particularly be considered. From an institutional point of view, one can say that the nation-state has indeed played a key role in this process of Westernisation. In its most recent stage (i.e. globalisation), however, the nation-state's importance somewhat declines, as other actors increasingly also come to play a role. We think here in particular of international and especially multilateral organisations, but also of transnational corporations (TNCs) and to a lesser extent of globally organised non-governmental organisations (NGOs). We will come back to this institutional dimension of globalisation in the next section of this chapter, as it is particularly relevant when it comes to understanding the transformation of water and other infrastructures in a globalised world.

However, before doing so, let us recall the two other key dimensions of this process of globalisation, namely the technological and the economic dimensions, both of which play equally important roles in recent infrastructure development in general, and in the water sector in particular.

The technological and economic dimensions of globalisation

Both technology and economics do play a key role in the process of globalisation, and thus deserve particular attention here. Indeed, *technology*'s role in the process of industrial development and modernisation is often underestimated, even though it is technology, which has made Western expansion possible, and has shaped corresponding cultural values. Infrastructure technologies, moreover, have played a particularly important role when it comes to modernisation and Westernisation, as they have significantly contributed to developing national territories. Not surprisingly, and until very recently, infrastructure technologies – be they in communications, transport, energy or water – have been closely tied to the role of the nation-state. More recently, however, and mainly thanks to the new information technologies, technological development seems to have emancipated itself to a certain extent from the nation-state and taken on global dimensions. As such, these information technologies are now all significantly affecting the traditional infrastructure technologies, either by connecting them into increasingly global networks (such as, for example, telecommunications or logistics networks), or by duplicating them virtually, and thus again connecting them at a more global level. In short, and thanks to their potentially universalistic nature, technology in general and infrastructure technologies in particular have played a significant role in the overall process of globalisation. The new information technologies are still accelerating this process, yet at the same time actively contributing to creating a new global (technological) reality.

Economic globalisation is generally considered to be globalisation *per se*. As such, it defines the globalisation of markets, measured in terms of trade flow. Even if trade flows are still a very conservative measure – given the fact that nation-states are taken as the unit of reference for exchanging goods and services – such flows have increased more or less steadily ever since the industrial revolution, and are being further pushed ahead by new global actors interested in promoting trade. Consequently, national economies are growing ever more dependent upon each other, a process which has been enhanced in recent years by so-called foreign direct investment (FDI). Indeed, FDI is now a key indicator of globalisation, and has, since the mid-1980s, grown faster than world output and trade (e.g. UNCTC, 1991). Hence, economic interdependence is increasingly production-based, and no longer just trade-based (UNCTAD, 1993). FDI, along with technological innovations, thus make up the main factors of this globalisation process. Infrastructures, as we will see later on, actually constitute a key proportion of such FDI, especially in developing countries, thus contributing even more to economic globalisation, while at the same time withdrawing infrastructures from state control.

Conclusion

We can thus see globalisation as a process by which decision-making power is gradually removed from the nation-states and shifted to other actors, which can be located 'above', 'across' and 'besides' the nation-states. We are thus basically interested in the organisational and institutional dimensions of the globalisation process, and seek to better understand what such globalisation means for solving collective problems, particularly in the case of infrastructures and more precisely in the case of water. As such, we want to explore what technological and economic globalisation in particular and globalisation in general mean for the traditional actors, such as the nation-state, and for their necessary transformation, as well as for the emergence of new actors such as TNCs or global NGOs. Indeed, some authors believe that globalisation is just another word for 'transnationalisation', i.e. the unfettered expansion of TNCs into the world economy, particularly into the economies of the developing countries (e.g. Raghavan, 1997a: 11). While we think that transnationalisation must be more broadly conceived and include in particular also global NGOs and new multilateral and international organisations, we nevertheless agree with Raghavan and many others that globalisation should be analysed, above all, in terms of globally operating actors and their dynamics. This, at least, is the approach we want to pursue in this book. Consequently and as a next step, we now have to clarify the organisational and institutional dimensions of globalisation.

The organisational and institutional dimensions of globalisation

The organisational and institutional dynamics of globalisation – themselves grounded in technological, economic, cultural and other changes – thus significantly affect the dynamics of the actors involved in solving collective problems, be it at a global level or elsewhere. In this section, we want to explore the implications of globalisation for the international organisations and the international system more generally, for the emergence of new transnational actors, and for the traditional nation-state.

Implications for international organisations

After the Second World War the United Nations (UN) were set up with a triple mandate – peace, development and human rights. Peace and development become institutionalised in the various agencies and inter-governmental mechanisms of the UN. Numerous UN agencies have emerged and developed, a process which in turn called for more inter-agency co-ordination. From an organisational and institutional point of view, the evolution of the UN and related international organisations can thus be described as a process of growing bureaucratisation and complexity.

This process slowed down somewhat in the late 1980s and came to a halt in the 1990s as a result of the various consequences of globalisation for the nation-states (see below), as well as for the entire UN system. Like all other public organisations, the UN and related agencies came under financial as well as legitimation pressures, a process which was accompanied by the emergence of new and equally powerful – if not more efficient and effective – global actors, such as TNCs and (global) NGOs. Consequently, one can observe, at the international level, the emergence of new concepts such as (global) 'governance' or 'partnerships', meaning that these new actors are now included in the management of international – or, rather, global – problems. So-called 'Summits' (e.g. Earth Summit, Social Summit, etc.) were often the occasion to include these new actors in the management of selected collective problems.

The United Nations Conference for Trade and Development (UNCTAD) Summit 'Partners for Development', held in Lyon in November 1998, may serve as an illustration of these new developments in and across international organisations. In a keynote address, its Secretary General, Mr Rubens Ricupero, spoke of civil society and the private sector as the real actors of development, thus implicitly downgrading the various UN agencies to a catalytic and facilitating role along with their 'partners'. He added that by working with the private sector, his organisation would demonstrate that profit and development, market and solidarity, were no longer in opposition but rather complementary. The central premise of the conference was that strategic partnership between governments and the private sector could play a significant role in promoting development. Just like UNCTAD, the United Nations Development Program (UNDP) has refocused its mission from a 'development' agency to a 'facilitator' of a development process which is basically carried forward by others, when it says in its new mission statement: 'The organisation assists in building capacity for good governance, popular participation, private and public sector development and growth with equity.'[2] More generally, most UN agencies have engaged in a similar change of view, and the UN as a whole has established a so-called 'Partners in Action Program', whose aim is to achieve UN objectives by means of partnerships, in particular with the private sector, i.e. TNCs.

But as a result of such globalisation, the UN and its different agencies not only undergo a transformation in terms of the way they see their role in the international arena (i.e. from advising governments and being actively involved in technical cooperation to facilitation, partnerships, governance, and network management, e.g. Bislev *et al.*, 2000). They also undergo substantial organisational transformation along new public management principles. As such, and under similar financial and legitimation pressures, they seek to become more efficient, as well as more effective, which is not always easy considering the heavy weight of bureacracy under which such international organisations operate. Yet, most importantly, the various international organisations are starting to think strategically as result of limited funds and competing actors. From having been structured along the ideas of peace and development, we now see the emergence of a new way of structuring international affairs along three different lines, i.e. the traditional development agenda, peace and security, as well as trade and corresponding regulation. UNCTAD and UNDP are perfect examples of this realignment, as they move away from the traditional UN and seek to align themselves with newly emerging powerful international actors, in the World Trade Organisation (WTO) and the Bretton Woods Institutions, respectively.

Indeed, we can observe three newly emerging poles among international organisations. The first pole is constituted around the WTO and pertains to issues of trade liberalisation and regulation, as trade and its regulation become one of the key issues of a globalised economy. UN agencies such as UNCTAD, but also NGOs such as the International Organisation for Standardisation (ISO) are regrouping around the WTO and the issue of trade. Thus, UNCTAD, for example, is no longer searching for a new development

2 UNDP mission statement <http://www.undp.org>.

paradigm or a new global economic order, but rather for policy guidelines which will help developing countries to adapt to the liberalisation of world trade, i.e. to allow them to participate more fully and take advantage of economic globalisation.

A second pole is constituted around what is left of the UN and pertains to issues of security and humanitarian affairs. Security and humanitarian affairs is indeed where the UN still has a competitive advantage and role to play in the international system. The trend here will be towards emergency operations, which will be mainly driven by security concerns. UNDP, for example, which was traditionally part of the core UN system, is thus increasingly moving away from the UN and towards the Bretton Woods Institutions. These institutions – in particular the World Bank and the International Monetary Fund (IMF) – constitute the third pole of the reorganisation of the international system.

Indeed, the World Bank in particular seems to be becoming the focal point around which issues of development, which were formerly part of the UN, now crystallise. More independent, as well as financially better off than the other UN organisations, the World Bank has managed to become the key international organisation in the area of development. As such, it has expanded its role – in particular thanks to Structural Adjustment Policies (SAPs) – from financing infrastructure projects to reforming entire states in order to promote industrial type development. The 1997 World Development Report, *The State in a Changing World*, illustrates this new perspective. For example, in the area of public services provision, this report emphasises that:

> Building an effective public sector . . . will mean opening up core government institutions, to improve incentives in areas that the public sector has long monopolized . . . contracting out services through competitive bids and auctions . . . in industrial countries . . . [and] in developing countries. Faced with weak administrative capacity, some countries . . . are also contracting out the delivery of social services to nongovernmental organisations.
>
> (World Bank, 1997c: 9)

In short, globalisation has had a significant impact upon international actors and the international system more generally. Not only has it forced these actors to modernise and to increasingly work in partnerships, especially with private actors, but it has also led them to regroup themselves around the three core issues of security (the UN), trade (the WTO) and development (the World Bank). In this book, we are particularly concerned with the third core issue of development, and thus with the role the World Bank now plays in a globalised economy (see Chapter 3).

Implications for transnational actors

Globalisation, as noted above, gives rise to new global actors, in particular TNCs and NGOs. These are the type of actors, which become partners of international organisations and, together with these organisations, as well as with nation-states, form new governance mechanisms. In this section, we will briefly present these new actors, and highlight how TNCs in particular, thanks to the process of globalisation, have acquired considerable clout and power, a fact, which is particularly relevant in the water sector.

The rise of NGOs remains a largely unexplored phenomenon, as NGOs have so far been approached essentially from a social action or social movement perspective, rather than from an organisational or institutional one (e.g. Princen and Finger, 1994). Yet, their numbers and their importance are growing. There are now said to be approximately 6000 NGOs worldwide (Union of International Associations, 2000: 2407), most of them, however, are active at a local level. Nevertheless, there is also an increasing number of international or global NGOs, especially in the environmental arena. Originally, NGOs in general and international NGOs in particular were mainly considered to be the representatives of civil society. At the international level, they were often seen as a counter-power to governments and TNCs (Lipschutz, 1996). However, we think that their role is gradually changing, as their democratic legitimation is increasingly being contested and as they are rapidly being integrated into the international system. From being a counter-power, they are now becoming partners in newly emerging global governance mechanisms. This evolution of NGOs is clearly observable in the water sector, where big international NGOs have emerged which present themselves as 'umbrella organisations' contributing to solving the global water problems. These will be explored in Chapter 2.

The other newly emerging actors in this process of globalisation are TNCs. The term 'TNC' seems to have emerged during the 1980s in order to describe an evolution among corporations active in the international arena (Levitt, 1985): from international corporations in the 1950s and 1960s, there was an evolution towards multinational corporations in the 1970s and 1980s, and more recently towards TNCs. This evolution must mainly be seen against the background of both economic and financial globalisation, as well as technological evolution: from standardised mass production the firms gradually moved into flexible production methods, whereby parts could be produced in different locations on the globe and assembled somewhere else. Consequently, these firms are no longer strong national monopolies, but acquire the form of production networks, all linked together by global communication and logistic systems. In short, TNCs are the typical product of economic, financial, technological but also cultural globalisations.

Depending on how TNCs are counted, there are now huge numbers of them. In 1993, for example, 40,000 such TNCs were counted (UNCTAD, 1993), and more recently up to 63,000 (UNCTAD, 2000: 9). In 1993, 70 per cent of the world's trade activities were said to be related to TNCs, and approximately half of this trade was simply intra-firm, i.e. trade activities that were internal to TNCs (UNCTAD, 1993). In 1999, FDI inflows by TNCs had reached the impressive level of $865 billion (UNCTAD, 2000: 3). FDI, as we have seen above, is a good measure of economic globalisation, and one can thus conclude that TNCs are therefore also a significant driver of such globalisation. As we will see below, FDI plays a particularly important role in the case of infrastructure development, especially in developing countries. In short, while globalisation gives rise to TNCs, they have now grown to the point where they are also actively driving the process.

But TNCs' growth is not solely the result of economic expansion: it also results from mergers and acquisitions (M&As). Indeed, there was significant M&A activity among TNCs throughout the 1990s. During that decade, there were mergers in most sectors but especially in telecommunications, banking, insurance, aviation and energy, reaching unprecedented volumes (UNCTAD, 2000, Chapter IV). From being limited to the United States, Europe and Japan, acquisitions in particular have now spread worldwide, in particular into developing countries in Latin America and transition economies in Eastern Europe. As we will see below in the case of the water sector, acquisitions are particularly prominent in the infrastructure sectors, given the fact that many developing countries, under the World Bank's SAPs, are now privatising their infrastructures. Acquisition of such privatised infrastructures makes up approximately $68.5 billion (UNCTAD, 2000: 106).

The consequences of such M&A activity are quite easy to imagine. Indeed, we can observe powerful concentration processes in all industry sectors, even in the most dynamic ones such as software and computer production. In the area of consumer durables, for example, 70 per cent of the market is already controlled by five firms. In the automotive, aviation and aerospace industries, the top five firms control around 50 per cent of the market, which must be considered as highly monopolistic (UNCTAD, 2000, Chapter IV). The infrastructure industries are not exempt from this evolution, and there are similar concentration processes in telecommunications, energy and water and sanitation. So far, such concentration has accelerated, rather than slowed down.

Not surprisingly, some TNCs have acquired huge economic – and, consequently, also political – power. Indeed, many TNCs have more cash reserves than governments (Anderson and Cavanagh, 2000). Among the 100 biggest economies, for example, one can already count a majority of TNCs (Korten, 1996). Quite logically, many TNCs have more power than governments, not only in bargaining processes (e.g. Rund, 1988), but also in setting political agendas and defining public policies, a fact Susan Strange has termed 'structural power' (Strange, 1988). TNCs, including water TNCs as we shall see, now have the ability to influence the formulation, as well as the implementation, of development strategies. This is especially the case when TNCs simultaneously lobby governments, especially Third World governments, and the World Bank as the main lender agency.

However, this political power of TNCs throughout all sectors is largely unexplored. In this book, we will highlight the power of some selected TNCs in the water sector, especially in relationship to state transformation and privatisation on the one hand and World Bank policy on the other. Even though we do not aspire to making a contribution in this regard, it must nevertheless be mentioned that the political power of TNCs remains also intellectually and theoretically largely unexplored. Of course, Susan Strange is well known for her efforts to focus attention on the political significance of firms as part of a more general shift from a politics based on states to politics based on markets and triangular diplomacy between states, between states and firms and between firms. She has indeed convincingly argued, that 'the progressive integration of the world economy, through international production, has shifted the balance of power away from states and toward world markets. This shift has led to the transfer of some power in relation to civil society from territorial states to nonterritorial TNCs; it has "actually made political players of the TNCs"' (Strange, 1996: 45–6). Strange furthermore argues that transnational corporations are now 'exercising a parallel authority alongside governments in matters of economic management affecting the location of industry and investment, the direction of technological innovation, the management of labour relations and the fiscal extraction of surplus value' (Strange, 1996: 65). But most of the literature on the subject has been written by activists, often with a Third World perspective.[3] Dr. Mahatir Mohamad, the Prime Minister of Malaysia, and a spokesperson for developing countries has, for example, asserted that globalisation is merely a new form of re-colonisation, this time by means of TNCs.[4] On the other hand, business theory still addresses the way firms react to national policies and actions or the way they shape government positions (e.g. Grosse and Behrman, 1992). And though international business theories in their analysis of TNCs often do address ownership specific advantages, they do so mainly in financial terms and in a strictly economic approach (e.g. Dunning, 1993). The respective power and strategic interests of TNCs in respect to governments and international organisations, however, still largely remain taboo or simply unexplored.

3 See for instance journals or organisations such as *Third World Resurgence*, *South–North Development Monitor*, *Third World Network*, *ATTAC*, *Multinational Monitor*, *Monde Diplomatique*, and many others.

4 Speech delivered by Dr Mahatir Bin Mohamad at the Inaugural Lecture of the Prime Ministers of Malaysia Fellowship Exchange Programme held in Kuala Lumpur on 24 July 1996, quoted in 'Globalisation: What it means to small nations', *Third World Resurgence*, 74 (1997), pp. 27–30. See also Raghavan (1997b).

Thus, most analysts seem to ignore the fact that TNCs are not only becoming major players in the global political economic system, but that they are now in competition with nation-states in areas which previously were exclusively in the hands of the state. More precisely, the relationship between TNCs and states, in the age of globalisation, seems to have become a strategic one. And this strategic relationship with the state is reinforced by the fact that TNCs can now also play a significant role in the international political scene by means of partnerships and other governance mechanisms. This same evolution can also be observed in the infrastructure, and especially the water sectors, where TNCs, as we will show, have already acquired significant power over states thanks in particular to their good international relations with the World Bank. However, the particular character of water infrastructures, which make it quite different from the traditional industry sector means that not only are they often monopolistic in nature, but they also deliver vital goods to people, which means that the state generally tends to be more directly involved. The power TNCs have over governments in these sectors is thus of even bigger concern. In order to understand this power, especially in the infrastructure sectors, we must however first understand how the state is itself changing as a result of globalisation.

Implications for the nation-state

We have seen that, at an international level and as a result of globalisation, the state is increasingly surrounded by new actors – from the World Bank to TNCs – with whom it will have to collaborate and work in partnership. On a national level, however, the effects of globalisation are less obvious, though, in our view, equally profound. In this section, we first want to briefly recall some historical elements of state development, then identify the three main pressures from globalisation upon the state, and finally see what this means in terms of current state transformation. Again, we look at the state from an organisational and institutional perspective, which means that we see it as an entity – or a conglomerate of entities – seeking to adjust and to survive in light of the new pressures of globalisation.

Historically, the state has expanded its functions and importance (measured in terms of expenditures) since its creation and until very recently (World Bank, 1997c: 2). From defence and security concerns in its very origins, the state – in the North and in the South – has gradually taken on so-called 'repair' activities in the health and social areas starting in the late nineteenth century and in the environmental area since the 1960s. As early as the late nineteenth century, but especially since the beginning of the twentieth century, the state has become particularly active in infrastructure development, thus financing, setting up and often also operating infrastructures in the transport, energy, telecommunications, postal, audiovisual, aviation, and of course water sectors. In many countries, especially Latin countries, the state has even become active in industrial production. Though not all these infrastructures were at the national level, corresponding policies, however, were mainly national ones. Not to mention of course the communist economies, where the state took on most productive functions. It is thus not surprising that not only the functions, but also the expenditures of the state have constantly grown since its inception. Where the state was not itself active in production, it intervened (and still does) in terms of macroeconomic policies.

> Thus by the 1960s the state appeared to be the dominant social entity: state and society were virtually coterminous. . . . Industrial states, East and West, were ramified public services agencies, omnicompetent to supervise and to provide for every aspects of the life of their communities . . . This perception of state has changed out of all recognition and with surprising rapidity. The revolutions of 1989 in Eastern Europe and their aftermath have led to a widespread perception of the modern world as one in which nation-states are losing their capacities for governance and national-level processes are ceding their primacy to global ones.
>
> (Hirst and Thomson, 1996: 174–5)

We can identify three types of pressures, all linked to globalisation, which have, since the late 1980s, started to put considerable pressure onto the state and especially onto its productive activities, many of them in the infrastructure sectors. These are financial, legitimation and ideological/political pressures. To start with the third, one must note that with financial and economic globalisation and the corresponding opening up of trade and deregulation, the state is somewhat weakened in its ability to conduct macroeconomic policy. Also, in many sectors, especially in infrastructures, emerging competition obliged and still obliges the state to loosen its control over its own public operators, in order for them to be treated like any other operator, as well as to be able to compete on the world market. As Dunning and Hamdani say:

> The second cause of globalization – which in many ways is better described as a removal of an obstacle – is the renaissance of market-oriented policies pursued by national governments and regional authorities. In the past five years alone, while more than 30 countries have abandoned central planning as the main mode of allocating scarce resources, over 80 countries have liberalized their inward FDI policies. The privatization of state-owned enterprises, the liberalization and deregulation of markets – especially for services – and the removal of a bevy of structural distortions, have all worked to stimulate cross-border corporate integration, both within TNCs and between interdependent firms or groups of firms.
>
> (Dunning and Hamdani, 1997: 15)

But such global market deregulation and corresponding reduction of the role of the state is itself the result of new neo-liberal economic theories – or, as some would argue, ideologies. Opposed to the Keynesian theory, which had inspired most countries from the 1950s onwards, the liberal and especially the neo-liberal school of thought of the so-called 'Chicago boys', now openly criticise state intervention in the economic activities as being too costly and in any case counter-productive for the global competitiveness of a country (Friedman, 1953). Public Choice theory moreover managed to argue that regulation, or any form of state intervention for that matter, was ultimately driven by lobbies, and thus in the private, rather than in the public interest (e.g. Lane, 1985). New developments in economic theory finally had significant impacts on infrastructures, as they argued for unbundling of formerly integrated activities, thus separating production, from transport, and distribution (e.g. Demsetz, 1989), as well as for tendering and competitive bidding (e.g. Baumol, 1969).

But besides these political, ideological or theoretical arguments for a reduction of the state in a national economy, globalisation has also led to some very significant pressures on the state: indeed, the need to increasingly compete on an international level, as well as often high public debt, has exerted some strong financial pressures: states now have to reduce their debt, limit public expenditures or otherwise 'adjust' to the new global economy. Such financial pressure is indeed in our view one of the main, if not the main, force which has pushed the state to undergo substantial transformation since the late 1980s, both in the North and in the South. In our opinion, the other equally significant force of global proportions leading to the recent transformations of the state is the so-called 'legitimation crisis' (Habermas, 1980; Offe, 1984). Indeed, with an increasingly global economy and culture, the state is more and more challenged to legitimise itself in light of the newly emerging supra-national actors on the one hand and public pressure from below (often locally and regionally) on the other. Not to mention the fact that, in the eyes of the citizens, the state increasingly has had to prove its usefulness, given in particular its weakening importance in light of rising private actors. In short and in the age of globalisation, the state has come under serious financial and legitimation pressures, and moreover must affirm itself in an increasingly global economy which obliges the state to behave more or less like any other (private) entity subject to competition.

It is thus not surprising that we can observe since the late 1980s significant efforts by the different nation-states to adjust to this new global economy and to globalisation more generally. While in the North (i.e. in the OECD countries), this generally takes the form of so-called New Public Management (NPM) efforts, in the South it takes the form of IMF- and World Bank-driven so-called 'structural adjustment programs' (SAPs) (Balassa, 1981). Yet, the ultimate outcomes, as well as the key underlying principles, of both NPM and SAPs might well be the same. Indeed, NPM is mainly the effort, on behalf of nation-states, to streamline administrative functions by focusing on service quality on the one hand, and to more clearly distinguish between politics and management on the other. Management – i.e. service delivery functions – are being increasingly outsourced and decentralised, if not privatised altogether. This is particularly the case in the infrastructure sectors, among which is water. In the South, as well as in transition economies, thanks to the pressures of the World Bank and the IMF, one proceeds often directly to privatisation. As in the North, political reform follows suit, though often more slowly than in the South. Underlying to such state reforms, as on the international level, are the ideas of partnership and participation, i.e. the ideas that the state should provide its services in partnership with private organisations and NGOs, while citizens should be participating more actively in service provision, either financially or in kind. So far, such structural adjustment, in both the North and the South, is basically seen, along neo-classical or neo-liberal economic ideas, as a process of reducing the importance of the state in society. In Chapter 7 of this book (pp. 210–33), we will show that this might well be an illusion, as the role of the state is of course being reduced when it comes to service delivery, yet being increased in terms of (public service) regulation.

But in any case, what we are witnessing here is a process by which nation-states, in both the North and the South, are structurally adjusting to various pressures stemming from financial, economic, technological and cultural globalisations. In doing so as organisations and institutions, states of course are trying to preserve as much power as possible. Where not otherwise possible, they give in and collaborate with more powerful actors than themselves, such as TNCs – with the danger, of course, that at some point they may be instrumentalised by the very organisations they partner with. This danger, we think, is particularly prevalent in the case of infrastructure services in general, and water in particular.

Implications for infrastructures

We now have all the ingredients to understand the evolution of infrastructures within this overall context of globalisation. Indeed, we have seen so far how nation-states are increasingly pressured to adjust to a global economic environment, how private actors (especially TNCs) have rapidly grown in the context of opening markets and how new international or rather multilateral institutions are restructuring the international arena in partnership with private actors. Let us now relate each of these three evolutions more directly to the developments in infrastructure.

Infrastructures – in telecommunications, postal services, transport, energy, audiovisual, aviation, and water or sewerage sectors – are among the first to come under pressure when the state has to readjust to the global economy. Indeed, in most countries, such infrastructure services were generally directly offered by the state, through so-called state-owned enterprises (SOEs), and sometimes through public enterprises belonging to local or regional authorities. If not directly offered, such infrastructure services were heavily regulated by the state. However, be it for ideological reasons as is the case in the United Kingdom, because of financial pressure, or simply because these SOEs needed to acquire managerial autonomy in order to compete with an increasingly open market, the state is gradually withdrawing from operating infrastructure services in all these sectors. This is, in most cases, but especially in the North, a slow process, whereby the private sector is gradually taking the place of the state as an operator, while the state, in theory, should retain – or even strengthen – its regulatory functions. The ultimate outcome of this process is probably the fact that such infrastructure services are going to be provided by the private sector, by companies with mixed public and private capital or by some other forms of public–private partnerships. There is no reason why this evolution should spare the water and sewerage sectors, as we shall see.

Globalisation, as we have seen, goes hand in hand with the growth of TNCs in all sectors whose markets have been opened up. This evolution can of course also be observed in the infrastructure sectors, where we see a similar growth and concentration process of TNCs. Table 1.1 lists the most important TNCs active in the infrastructure sector in relationship to the Fortune 500 companies. From it, one can see that 'infrastructure TNCs' already occupy a prominent role among TNCs in general. It is likely that the relative importance of such infrastructure TNCs will only increase parallel to further privatisation.

Table 1.1 Infrastructure TNCs within the first top 200 TNCs, 2000 (compiled from Fortune 500 list)

Global rank	*Company*	*Revenues ($ million)*	*Global rank*	*Company*	*Revenues ($ million)*
13	Nippon Telegraph & Telephone (NTT)	93,591.7	95	Suez Lyonnaise des Eaux	33,559.7
26	US Postal Services	62,726.0	97	Verizon Communications	33,174.0
28	AT&T	62,391.0			
37	E.ON	52,227.7	118	France Télécom	29,048.8
48	Tokyo Electric Power	45,727.7	130	United Parcel Service	27,052.0
			156	BellSouth	25,224.0
53	Vivendi	44,397.8	157	Deutsche Post	25,101.1
62	Enron	40,112.0	164	Telefónica	24,487.7
73	RWE Group	38,357.5	178	Kansai Electric Power	23,246.2
77	Deutsche Telekom	37,835.1			
79	WorldCom	37,120.0	183	East Japan Railway	22,478.5
83	State Power Corporations	36,076.1	186	Enel	22,320.1
			190	Duke Energy	21,742.0
88	Electricité de France	34,146.6	194	Nortel Networks	21,287.0
89	ENI	34,091.0	200	Viag	20,758

And, indeed, privatisation seems to be the main policy of the World Bank, the major international actor in the infrastructure sector. As we have seen, the World Bank, along with the IMF, emerge in the age of globalisation as the actors in charge of development, and thus also in charge of infrastructure development. Chapter 3 will show the World Bank's particular view on how development should occur in the water sector. Let us simply state here that, for the World Bank, development is in essence a process, which should occur in partnership, especially in partnership with the private sector. Even privatisation, for the World Bank, seems to be a partnership. Consequently, the World Bank advocates various forms of public–private partnerships for infrastructure maintenance and development, as shown in Table 1.2.

If not operated by the private sector, infrastructures should be run, according to the World Bank, like a business, as highlighted in Box 1.1 below.

It appears from Tables 1.1 and 1.2 that what counts is the management of infrastructures following private sector and market principles, and that privatisation is just one of the options. For example, it was estimated that between 1990 and 1999, nearly $580 billion had been invested by private

Box 1.1 The World Bank's view on how to manage infrastructure services

- **Manage infrastructure like a business, not a bureaucracy.** The provision of infrastructure needs to be conceived and run as a service industry that responds to customer demand. Poor performers typically have a confusion of objectives, little financial autonomy or financial discipline, and 'no bottom line' measured by customer satisfaction. The high willingness to pay for most infrastructure services, even by the poor, provides greater opportunity for user charges. Private sector involvement in management, financing, or ownership will in most cases be needed to ensure a commercial orientation in infrastructure.
- **Introduce competition, directly if feasible, indirectly if not.** Competition gives consumers choices for better meeting their demands and puts pressure on suppliers to be efficient and accountable to users. Competition can be introduced directly, by liberalising entry into activities that have no technological barriers, and indirectly, through competitive bidding for the right to provide exclusive services where natural monopoly conditions exist and by liberalising the supply of services substitutes.
- **Give users and other stakeholders a strong voice and real responsibility.** Where infrastructure activities involve important external effects, for good or bad, or where market discipline is insufficient to ensure accountability to users and other affected groups, governments need to address their concerns through other means. Users and other stakeholders should be represented in the planning and regulation of infrastructure services, and in some cases they should take majors initiatives in design, operation, and financing.

Source: World Bank (1994d: 2, Box 2).

Table 1.2 New forms of private sector participation in infrastructure maintenance and development

Options	*Property*	*Operations*	*Investments*	*Risks*	*Length (years)*
Service contract	Public	Public and private	Public	Public	1–2
Management contract	Public	**Private**	Public	Public	3–5
Leasing	Public	**Private**	Public	Shared	8–15
BOT		**Private**	**Private**	**Private**	20–30
Concession	Public	**Private**	**Private**	**Private**	25–30
Privatisation	**Private**	**Private**	**Private**	**Private**	undefined

Source: World Bank (1997a: 90, Table 2).

companies in developing countries, most of which was in the telecommunications and gas/electricity sectors (Izaguirre and Rao, 2000). Between 1990 and 1999, around 1900 infrastructure programmes had taken place in the developing countries, most of them BOT (Build–Operate–Transfer) programmes (Izaguirre and Rao, 2000). Nevertheless, the overall tendency here appears to be the privatisation of infrastructures, at least in the South where World Bank policies do have a significant impact. Table 1.3 highlights the importance of infrastructure privatisation in developing countries. This table also shows the relative place that water privatisation occupies in the overall privatisation phenomenon.

All implications of globalisation for infrastructure thus seem to converge around the same result – i.e. privatisation. Indeed, the state's role in providing such services has particularly come under attack in terms of operation, maintenance and investments, and especially in developing countries and transition economies. TNCs seem particularly keen to take on such infrastructure services, be it in terms of management, maintenance, investment or full operation. And the World Bank as the new global actor in infrastructure development appears to encourage this evolution.

This book will focus on water management in the age of globalisation. Yet, the above trends in the evolution of nation-states, transnational organisations and international public actors will constitute the background of such water management in both the North and the South. However, before we can fully understand what is going on in the water sector, we still need to examine the evolution in the environmental arena, as water is, after all, not only a good, but also an environmental problem and issue.

Table 1.3 Infrastructure privatisation in developing countries, 1990–9 ($US billions)

	1990	*1991*	*1992*	*1993*	*1994*	*1995*	*1996*	*1997*	*1998*	*1999*	*Total*
Sector											
Telecommunications	6.7	13.3	8.1	11.0	19.6	24.0	29.9	42.8	54.3	39.2	249.0
Energy	1.6	1.2	12.1	14.6	17.0	24.1	33.7	47.9	25.7	14.9	192.8
Transport	8.0	3.1	4.2	7.7	8.2	10.1	16.5	22.4	17.6	8.4	106.1
Water and sanitation	–	0.1	1.9	7.5	0.7	1.7	2.2	8.9	2.6	5.9	31.4
Region											
East Asia and Pacific	2.6	4.1	8.9	16.2	17.7	23.4	33.4	38.8	9.5	14.1	168.6
Europe and Central Asia	0.1	0.3	1.3	1.5	3.9	8.6	11.6	15.1	11.5	8.7	62.5
Latin America and the Caribbean	13.2	12.6	15.8	18.5	18.9	19.4	28.8	51.1	71.0	36.3	285.6
Middle East and North Africa	0.0	–	0.0	3.4	0.3	0.1	0.4	5.3	3.5	2.4	15.3
South Asia	0.3	0.8	0.1	1.3	4.0	7.6	6.1	7.1	2.3	4.0	33.5
Sub-Saharan Africa	0.1	–	0.1	0.0	0.7	0.8	2.1	4.5	2.4	2.9	13.6
Total	**16.3**	**17.8**	**26.1**	**40.9**	**45.5**	**59.9**	**82.3**	**121.9**	**110.2**	**68.5**	**579.3**

Note: 0.0 means zero or less than half the unit shown. Data may not sum to totals because of rounding.
Source: World Bank PPI Project database, quoted in Izaguirre and Rao (2000).

Chapter 2

Environmental pressures and the reorganisation of the water sector

> Recent international forums dealing with water resources issues have all brought about an increasing awareness of the global magnitude of a water crisis. This possibility was . . . brought to the attention of the world community at the time of the United Nations Water Conference in 1977, and renewed notes of alarm were sounded at the International Conference on Water and the Environment and at the United Nations Conference on Environment and Development. Although these expressions of concern may be deemed by many to be insufficient, there is a growing consensus among experts in the water resources field as to the seriousness of the situation. However, the spectre of a global water crisis has been overshadowed by concerns about other issues of manifest global proportions, such as ozone layer, tropical forests and climate change. Internationally the seriousness of water problems has not as yet received the recognition warranted [by] the situation.
>
> (United Nations Committee on Natural Resources (UNCNR), 1994: §73)

One of the major trends to have marked past decades is certainly environmental degradation. Such environmental degradation is of course strongly related to the much larger dynamics of industrial development and modernisation. As such, the overall concern with environmental issues and problems cannot be separated from a broader critical look at Western-style society and its expansion globally. In short, there exists, in our view, an ultimately unsolvable tension between further industrial development and expansion on the one hand and environmental protection on the other. Water not only illustrates, but moreover is at the core of this tension: water came into the global limelight primarily because of environmental concerns, yet at the same time water has become an object of technological and managerial solutions to solve the environmental problem so created. As we will see in this chapter, the major reforms initiated in the water sector all stem from increased environmental preoccupations. As with other sectors, such as climate change, the awareness of water problems came late and in some parts of the world the water problem cannot be brought back to a sustainable path.

The purpose of this chapter is to offer an understanding of how water is currently being approached by the different actors involved, focusing primarily on international actors. It is our assumption that the way these actors see the water problem significantly influences how water will be approached in the future. In order to develop this understanding, we proceed in four separate steps: in a first section, we will trace the history of the international concern about water, mainly to show that water became a concern in the wake of overall environmental problems. Quite logically, we will thus focus on the historical events and conferences that have catalysed the need for international co-operation among states, but also among international organisations, development agencies and water specialists. The so-called 'Dublin Principles' appear to us to be the cornerstone around which the different international approaches to solving the global water problems crystallise. Indeed, the Dublin principles today summarise the currently existing international consensus on water policies to be followed at international, regional, national and local level. As such, these Dublin principles are, in our view, the lowest common denominator the different actors can agree to, rather than a particularly original and innovative approach to a newly emerging global concern. In fact, the core message of the Dublin principles is quite simple and says that water needs to be better managed, especially along economic principles, if sustainable development is ever going to be reached.

Our second section will complement the first. In it we want to offer a more factual view of what the issues and problems in the area of water currently are. This section thus presents facts and figures, whose aim it is to highlight the urgency of the current water shortages and degradations. We furthermore want to contrast this factual urgency with the consensual discourse arising from the first section and culminating in the Dublin principles. This section thus highlights the tension between the water problems resulting from further industrial development on the one hand and the proposed economic, environmental and managerial approaches to solving these problems on the other.

Our third and fourth sections will then present the different non-governmental organisations (NGOs) involved in the global water sector. Indeed, our organisational approach requires a solid understanding of how the various involved actors see the problem and how they act to addressing them. Following the International Drinking Water Supply and Sanitation Decade (1981–90) and the international conferences of the early 1990s, three major institutions were created, expressing this overall sense of urgency, namely the Global Water Partnership (GWP), the World Water Council (WWC), and the World Commission on Dams (WCD). All three will be presented in the third section. Besides examining these newly created institutions, we will, in our final section, also discuss the major international (mainly governmental) organisations dealing with global water problems. We will thus see that the

UN and others approach water from a scientific, from a health, from a development and to a much lesser extent from an environmental perspective.

In short, this chapter will, along with Chapter 1, analyse the different external trends that have influenced water resources management practices, by focusing on the main actors in water resources management today.

The history of the growing international concern about water resources

Concern over freshwater problems has been on the global agenda since the United Nations Conference on the Human Environment in Stockholm in 1972. Of course, UNESCO had already been involved since the early 1950s in water resources issues. UNESCO actually developed the first programme on freshwater issues in semi-arid and arid regions. In fact, between the 1950s and 1970s, UNESCO gradually played a growing role, going far beyond purely technical and scientific aspects. Indeed, UNESCO has been one of the most active international organisations pushing the question to the top of the environmental agenda. But the United Nations Water Resource Conference at Mar del Plata in 1977, which focused entirely on freshwater resources, constituted the first real attempt by international organisations to alert the international community to the dangerous overuse of water resources and the increased water scarcity observed in many regions of the world. Not surprisingly, at that time, the conference was still part and parcel of the development paradigm (e.g. Chatterjee and Finger, 1994). Indeed, the main concern expressed during the conference was that 'all peoples, whatever their stage of development and their social and economic condition, have the right to have access to drinking water in quantities and of a quality equal to their basic needs'.[1]

As a result of the Mar del Plata Action Plan, the International Drinking Water Supply and Sanitation Decade was launched in 1981, aiming at providing safe drinking-water and sanitation to underserved urban and rural areas by the year 1990. However, this objective was far from being realistic and it was not until the 1990s that the matter was more seriously approached by the UN and the international community.

From our point of view, 1990 was indeed the turning point: at that moment, the international community realised that water could not be addressed within the traditional development paradigm (e.g. pumping more water and cleaning it), but that environmental factors had also to be considered in water resources management and development. Development agencies such as the Canadian International Development Agency, the French Ministry of Co-operation, the German Federal Ministry of Economic Cooperation and

1 United Nations Water Conference, Mar del Plata (1977), quoted in UNICEF (1995: 5).

Development, the Overseas Development Administration (UK), and the US Agency for International Development all started, at that time, to develop new water resource strategies for foreign assistance.[2]

Simultaneously, civil society started to manifest its growing concern about environment matters and focused, among others, on the newly emerging challenges in the water sector. The Aral Sea Crisis, for example, created a certain sense of urgency. In 1990, several international forums on water were held, such as for example, the Global Consultation on Safe Water and Sanitation for the 1990s in New Delhi. During this conference, the New Delhi Statement reiterated the need to provide, on a sustainable basis, access to safe water in sufficient quantities and proper sanitation for all, emphasising the 'some for all, rather than more for some' approach. More precisely, the Delhi Statement focused on the four following principles:

- Protection of the environment and safeguarding of health through the integrated management of water resources and liquid and solid wastes
- Institutional reforms promoting an integrated approach and including changes in procedures, attitudes and behaviour, as well as the full participation of women at all levels in sector institutions
- Community management of services, backed by measures to strengthen local institutions in implementing and sustaining water and sanitation programmes
- Sound financial practices, achieved through better management of existing assets, and widespread use of appropriate technologies.[3]

As it turns out, these four principles substantially shaped what became known later as the 'Dublin Principles' (see pp. 24–5).

During the same year, at the World Summit for Children in September 1990, heads of State called for both universal access to water supply and sanitation and the eradication of Guinea worm disease by 1995.[4] The third conference on water-related issues during the same year was the 1990 Montreal International Forum entitled 'NGOs Working Together'. Again, this Forum highlighted the problems of supplying drinking water and sanitation. More precisely, 'The delegates called for a new approach to managing water resources that emphasizes wide access to potable water and to sanitation, user participation and consultation, reliance on local community resources, the improvement and repair of existing systems, and a policy of

2 See for example DANIDA (1992), SIDA/IRC (1994), BMZ (1996), ODA (1996), Netherlands Development Assistance (1998).

3 From Agenda 21, Chapter 18, § 18.48, as adopted by the UN Conference on Environment and Development, Rio de Janeiro (June 1992) <http ://www.unep.org/Documents/>.

4 This objective was not attained. See Sharma *et al.*, 1996: 9.

comprehensive water resource management integrating environmental and economic considerations' (World Bank, 1993: 4).

The proliferation of such conferences, the growing interest of civil society and NGOs for water and water-related problems, as well as increased environmental preoccupation more generally certainly brought water to the top of the international agenda. All these concerns culminated in 1992 with the so-called 'Dublin Statement' at the International Conference on Water and the Environment. In this context, one must also mention Chapter 18 of Agenda 21, a document adopted at the 1992 UN Conference on Environment and Development in Rio de Janeiro. It is very important to note that, in the case of water, the environment emerges, in the early 1990s, as a basically new concern. So far, water and sanitation problems had been dealt with from a development and health perspective. The integration of the environmental factor into water issues brought therefore a great change in the way water resources protection and management would be approached in the future. This is to what we now turn, starting with the Dublin Principles.

Dublin as the turning point

It is indeed at the International Conference on Water and the Environment in Dublin and at the 1992 UN Conference on Environment and Development in Rio that new approaches for water resources management emerged. These approaches culminated in 1992 in the so-called four 'Dublin principles', which defined the basic guidelines for water resources management in the years to come. These principles are as follows:[5]

- Principle 1: Freshwater is a finite and vulnerable resource, essential to sustain life, development and the environment. This means that since water sustains life, effective management of water resources demands a holistic approach, linking social and economic development with protection of natural ecosystems. Effective management links land and water uses across the whole of a catchment area or the groundwater aquifer.
- Principle 2: Water development and management should be based on a participatory approach, involving users, planners and policy-makers at all levels. Such a participatory approach involves raising awareness of the importance of water among policy-makers and the general public. It also means that decisions are taken according to a principle of subsidiarity, with full public consultation and involvement of users in planning and implementation of water projects.
- Principle 3: Women play a central part in the provision, management

5 Quoted from the *International Conference on Water and the Environment: Development Issues for the 21st century* <http ://www.wmo.ch/web/homs/icwedece.html>.

and safeguarding of water. This pivotal role of women as providers and users of water and guardians of the living environment has seldom been reflected in institutional arrangements for the development and management of water resources. Acceptance and implementation of this principle requires positive policies to address women's specific needs and to equip and empower women to participate at all levels in water resources programs, including decision-making and implementation, in ways defined by them.

- Principle 4: Water has an economic value in all its competing uses and should be recognised as an economic good. Within this principle, it is vital to recognise first the basic right of all human beings to have access to clean water and sanitation at an affordable price. Past failure to recognise the economic value of water, it is argued, has led to wasteful and environmentally damaging uses of the resource. Managing water as an economic good is an important way of achieving efficient and equitable use, and of encouraging conservation and protection of water resources. Principle 4 is thus the main innovation, leading directly to an economic approach to water resources. It is also the most contested principle among the actors involved in water issues (Perry, *et al.*, 1997).

Several months later in June 1992 the UN Conference on Environment and Development in Rio de Janeiro, Brazil, reiterated the Dublin Principles when saying in particular that: 'Integrated water resources management is based on the perception of water as an integral part of the ecosystem, a natural resource and a social and economic good, whose quantity and quality determines the nature of utilization'.[6] Furthermore, this conference also stressed the fact that 'the implementation of allocation decisions through demand management, pricing mechanisms, and regulatory measures'.[7]

In short, both Dublin and Rio represent a fundamentally new way of thinking, as, from now on, environmental issues are clearly placed before and above development concerns. More precisely, the underlying ideas of both Dublin and Rio can be summarised in the following seven points:

- Environment and the protection and conservation of freshwater supply are key
- Available water resources must be monitored
- The allocation of water resources must be optimised, which raises the issue of irrigation

6 From Agenda 21, Chapter 18, §§ 18.6 and 18.8, as adopted by the UN Conference on Environment and Development, Rio de Janeiro (June 1992) <http ://www.unep.org/Documents/>.

7 From Agenda 21, Chapter 18, § 18.12 (f), as adopted by the UN Conference on Environment and Development, Rio de Janeiro (June 1992) <http ://www.unep.org/Documents/>.

- Water is an economic good and must be managed according to demand
- Technology is paramount
- Decentralisation and user participation must be encouraged
- Regional co-operation and international awareness must be promoted.

As we will see throughout this book, these seven points will recur throughout our analysis in the different cases and organisations studied. Underlying these different points are actually three separate perspectives – i.e. a political, an economic and an environmental perspective. However, it is the economic perspective, which takes precedence over the others. Indeed, the Dublin Principles reflect a combination of all three.

These Dublin Principles are in fact so pervasive that the main authors in the field, and even the most critical ones, have been obliged to acknowledge them. Let us thus briefly discuss here the views of Ricardo Petrella, one of the most critical persons in the field, of James Winpenney and of Sandra Postel from the Worldwatch Institute.

Selected authors' thoughts on the global water crisis

Our first example is Ricardo Petrella's initiative for a world water contract (Petrella, 1999). This initiative results from a simple fact: since 1977, as we have already stressed above, there has been a growing concern with water resources management. Several billion dollars have been invested in different international organisations' projects to provide water and sanitation for all. However, despite these efforts, more and more people do not have access to water (Petrella, 1999: 20). There is therefore, from Petrella's point of view, a need to change the way water is currently being managed. The environmental dimension is, in his view, one of the main problems, if not the main problem and should be addressed primarily in any reform effort. However, Petrella is one of the main opponents of recognising water as an economic good as defined according to the Dublin Principles. Petrella considers that instead of an economic good, water should be considered as a public good. More precisely, it has to be considered as 'un bien commun patrimonial mondial de l'humanité' – i.e. a common good belonging to humanity as a whole (Petrella, 1999: 19). Nevertheless, his analysis also leads him to conclude that it is the state which is mainly responsible for this mismanagement of the water resources, due, in particular, to corruption and highly centralised rigid structures. His proposed solution therefore consists of a 'désétatisation de l'Etat', i.e. a disengagement of the nation-state from the sector (Petrella, 1999: 27). As emphasised in the Dublin Principles, Petrella advocates decentralisation, user participation and a 'citizens' state'. However, such disengagement does not mean privatisation, but rather 'handing over of the management of water resources to local communities, citizens' groups, co-operatives, cities, and regional networks' (Petrella, 1999: 29).[8]

At the other extreme of the 'water spectrum' we can find James Winpenny, a former research fellow at the Overseas Development Institute (ODI) in the United Kingdom. He develops another theory of water resources management, this time from an economic and not from an activist perspective. Winpenny's first affirmation is that water must be recognised as a limited resource. In his view, the fact that this has not been done has created conflicts, environmental catastrophes and governments' failure to provide water for all. Indeed, Winpenny considers that water, which is already a scarce resource, should not be treated as though it were plentiful and free. Mainly responsible for this mismanagement are, according to him, agriculture, industry and the 'politicisation' of water allocation. The solution is therefore to measure the real cost of water. This will, Winpenny argues, enable us to increase water supply coverage, as well as encourage private investments. He says: 'low prices depress the profitably of investment in the water sector' (Winpenny, 1994: 18). He sees the enabling environment for such a policy in a clear separation between the operator and the regulator, as well as in measures that provide for real costs, water tariffs and demand management. The key focus of such water resources management must, according to Winpenny, be on the demand side. Consequently, Winpenny relies essentially on market mechanisms at the local level. In his final analysis, Winpenny concludes that the solution to the current water problems must be solidly grounded in treating water as an economic good, as an economic approach ultimately brings both financial and environmental benefits.

Between the political position of Petrella and the economic perspective of Winpenny, one can locate the environmental and conservationist perspective of Sandra Postel from the Worldwatch Institute. She is indeed concerned by both efficiency and conservation when saying, for example:

> In a sense, masking scarcity is a principal aim of water development, the collection of engineering projects and technologies that give people access to and control over nature's supply. But all too often it has proceeded without regard for harmful side-effects. We build ever more and larger projects to meet spiraling demands wherever they arise, but pay little mind to the ecological services of rivers, lakes, and wetlands that are lost in the process.
>
> (Postel, 1992: 18)

8 Authors' translation (à des entités publiques telles que des communautés locales, des groupements de citoyens, des réseaux de villages ou de villes, des sociétés coopératives).

According to her, the growing pressures on water resources thus call for a new approach to water resources management. Consequently, she sees the following main challenge: 'Achieving water balance will not be easy. The policies, laws and practices that shape water use today rarely promote all three basic tenets of sustainable resource use – efficiency, equity, and ecological integrity' (Postel, 1992: 22). As a matter of fact, the author follows the logic of the first Dublin Principle when she says that although water is a renewable resource, it is also a finite one. Postel quite logically proposes more investments in new technologies, especially in irrigation, innovative methods in small-scale projects, increased wastewater treatment and industrial recycling. She says:

> Exploiting the market potential of new water-saving technologies is an obvious one. And in many cases, achieving better water management will require decentralizing control over water, and moving from top-down decision-making to greater people's participation – a shift necessary for better human and economic development overall.
>
> (Postel, 1992: 24)

Water pricing, the elimination of government subsidies especially for irrigation, and the active role of women, are also factors that can achieve both efficiency and conservation. Sandra Postel's theory therefore relies on an environmental management approach while at the same time emphasising the need for a 'water ethics': 'Adopting such an ethic would represent a historic philosophical shift away from the strictly utilitarian, divide-and-conquer approach to water management and towards an integrated holistic approach that views people and water as related parts of a greater whole' (Postel, 1992: 185).

With these three authors, we think, the main three approaches to dealing with water have been identified, namely a political (Petrella), an economic (Winpenny) and an environmental (Postel) approach. But for all three authors the environmental concern is the underlying one. For all three, also, the role of the State in the water sector needs to be re-examined, considering, in fact, partnerships or other more innovative structures. It is also interesting to note that all three authors implicitly or explicitly refer to the Dublin Principles, which is not surprising given the fact they are quite broad and fuzzy. Yet, we still need to understand the underlying assumptions of these Dublin Principles a little better.

The essence of Dublin: an environmental crisis

From our point of view, the Dublin principles, as well as Agenda 21, are the result of a perceived urgency, a 'water crisis', rather than an international consensus. Indeed, most countries are now suffering from water stress. In 1994, for example, a Committee on Natural Resources of the Economic and Social Council reported to the Secretary General of the UN that:

> While in the past there was a tendency to regard water problems as being local or regional in nature, there is now a growing recognition that their increasingly widespread occurrence is quickly adding up to a crisis of global magnitude . . . water scarcity relative to demand is no longer a problem in arid or semi-arid areas alone, but it is now a common occurrence in both developed and developing countries alike.
>
> (UNCNR, 1994: § 9)

Like many general statements of this kind, the Dublin Principles had the virtue of being sufficiently vague in order to permit an agreement, while leaving the operational contents – over which there might be strong disagreements – unspecified. It is this urgency, along with the desire not to criticise any particular country or region, which leads to recommending some generally applicable principles on global water management, in particular Principle 4 (i.e. to consider water as an economic resource). In doing so, Dublin (and Rio) do not raise the issue of water management and corresponding institutional approaches, but rather seek harmonisation by using market mechanisms. In other words, for us, jumping immediately to Principle 4 – which is made possible thanks to the perceived urgency – makes a significant shortcut by ignoring, among others, the issue of (public) management, organisations and institutions. In short, economics is supposed to find a solution to a problem which, in our view, is essentially managerial, organisational, institutional and ultimately political in nature.

There are significant differences between the continents when it comes to water management. In Africa, for example, the situation is very critical. Few countries have services which can be compared favourably with those existing ten or twenty years ago. Latin America, as well as Asia and the Pacific, continue to face considerable problems. In addition, conditions in Eastern Europe and successor States of the former Union of Soviet Socialist Republic (USSR) have recently worsened, to the extent that many nations in Eastern Europe and Central Asia have suffered serious reductions in their hydrological services. In terms of the reforms proposed since the Del Plata Action Plan, there seems to be a significant discrepancy among the different continents. In the 1994 United Nations Review of Progress on water-related issues it was noted that:

> Although the Economic Commission for Africa has not recently conducted a survey of institutional responses to the recommendations of the Mar Del Plata Action Plan and of Agenda 21, the view is expressed that in the light of serious political and economic problems, little if any significant success is evident in this regard, and that the fragmentation of institutional responsibilities hinders the formulation of holistic approaches for the integrated development of water resources.
>
> (UNCNR, 1994: § 40)

In Asia and the Pacific, considerable progress has been made in the formulation of national water policies. However, it was noted that although floods were a major concern in this region, most countries had no adequate structural or non-structural measures to deal with this problem. In the Latin American region, according to the Economic Commission for Latin America and the Caribbean (ECLAC), many issues inherent in water system operation are being poorly handled and even ignored. Most of these differences among these countries and regions stem from differences in water management, not from failures in economic policy.

This managerial nature of the current water problems is enhanced by the fact that we are faced, in the water sector, with an unsolvable dilemma: indeed, while there is now a widespread consensus that actions and innovative approaches need to be taken so as to better manage water resources because of growing scarcity, there is also the widespread concern that all peoples should have access to water supply and sanitation. This duality corresponds to what Serageldin (1994) has called the co-existence of the 'old agenda' – i.e. to provide household services – with the 'new agenda' – i.e. to promote environmentally sustainable development. It is difficult to see how declaring water as an economic good will get us out of this dilemma, which is profoundly political in nature.

In the 1995 Copenhagen Declaration on Social Development, governments were urged to comply with their commitments towards promoting and achieving the highest attainable standards of physical and mental health and access to primary health care, including the developing and updating of country action plans or programmes to ensure universal, non-discriminatory access to basic health services, including sanitation and drinking water.[9] Similarly, the so-called 'Water Social Charter' elaborated by the Académie de l'Eau (Water Academy) in collaboration with the World Water Council and the World Commission on Water for the 21st century (Valiron and Roche, 2000), as well as the Earth Charter (Sweden), the Health Charter (United Kingdom), the OECD Mader Declaration, the Water Ethics Report by UNESCO and the World Water Contract by the Lisbon Group are generally endorsed by governments, thus reflecting their willingness to ensure that everybody should have access to water. And during Habitat II in 1997, the Istanbul Declaration on Human Settlements by the heads of states or governments endorsed once more the concerns and priorities set out in the different UN Conferences on Environment.[10]

However, one should not be misled by this 'double speak'. Indeed, while governments keep up the same political rhetoric as in the previous decades, Serageldin's 'old agenda' seems to be increasingly forgotten – i.e. the

9 World Summit for Social Development (Copenhagen, 6–12 March 1995, Chap. I, Resolution 1, Annex 1, Sect. C, commitment 6 (m)).

environmental preoccupation seems to have taken precedence over the development agenda, and the data in the next section will certainly show how environmental concern and corresponding environmental management are increasingly becoming the dominant approach. Moreover, governments should no longer be considered the main and only actor in water resources management and policies. Again, Dublin appears to be the moment where not only economics emerges as the solution, but also the relative importance of governments is reduced. Indeed, as stated in the report of the Secretary-General during the Eighth Session on Sustainable Development: 'While the role of governments remained as essential as ever, the nature of its intervention has undergone a considerable evolution. At the same time, other agents of progress have emerged as being critically important' (UNCSD, 2000: § 56).

Consequently – and as we have shown in Chapter 1 – since the mid-1990s, many more actors are now being involved in the water sector and new partnerships have emerged among them. As emphasised during Habitat II in the case of urban management:

> The urban environmental crisis will continue to be one of the most pressing problems facing humanity in the twenty-first century. Most of the world's gravest environmental threats to air quality, water quality and availability, waste disposal, and energy consumption are exacerbated by the high density and activity of urban life. Governments acting alone cannot successfully address these challenges – what is needed are partnerships between local governments, the private sector and citizens' groups working together to find solutions.[11]

In short, as we argue, Dublin is indeed the pivotal point where water is seen in terms of an environmental imperative. This in turn calls for an economic approach by which water is considered a scarce resource, and which now also has to be cared for by actors other than the State. Before turning to these

10 *The Habitat Agenda: Istanbul Declaration on Human Settlements* <http ://www.undp.org/un/habitat/agenda/ist-dec.html>. One should also mention the Ministerial Conference on Drinking Water in Noordwijk that recommended that governments undertake a water resources assessment in order to produce an inventory of the current situation and to identify problems and constraints in providing water supply and environmental sanitation services. The Conference also called on governments to develop, review or revise, in the context of a rational sustainable development strategy consistent with Agenda 21, measures for water resources management, including drinking water and environmental sanitation, and to develop, review or revise by 1997 and implement measures for achieving a rational and effective provision and use of drinking water and environmental sanitation (United Nations Commission on Sustainable Development (UNCSD), 2000: §79).

11 *Habitat II: Partnership for the urban environment*, United Nations Department of Public Information <http ://www.undp.org/un/habitat/presskit/dpi1790e.htm>.

actors, let us briefly look at the environmental imperative, i.e. the pressures and challenges in the water sector.

Pressures and challenges

Dublin is thus indicative of urgency rather than of consensus. In order to fully understand the forces, which have driven this shift towards a new environmental approach, one must briefly turn to the current reality in freshwater resources and the resulting challenges.

The importance of freshwater was highlighted in the fifth session of the Commission on Sustainable Development. It was defined as one of the most essential elements supporting human life, economic growth, and development. As recalled by this Commission, '[water] is irreplaceable for the purposes of drinking, hygiene, food production, fisheries, industry, hydropower generation, navigation, recreation and many other activities. Water is equally critical for the healthy functioning of nature, upon which human society is built' (UNCSD, 1997: § 32).

However, the world is facing increasing problems of water scarcity. Although 70 per cent of the world is covered with water, only 2.5 per cent is freshwater and human development has lead to a considerable increase in freshwater use (UNCSD, 1997: § 33). As emphasised by Peter Gleick, there have been three major driving forces in the enormous expansion of water resources infrastructure in the twentieth century, namely population growth, industrial development and the expansion of irrigated agriculture (Gleick, 1998). Indeed, all three factors have increased dramatically: between 1900 and 1995, the population of the world grew from 1.6 billion to nearly 6 billion people. Land under irrigation increased from around 50 million hectares in 1900 to over 250 million hectares in 2000. These and other factors have led to a nearly sevenfold increase in freshwater withdrawal.

At the same time, more than 1 billion people still lack access to safe drinking water, and about half of the world population lacks access to sanitation (e.g. Cosgrove and Rijsberman, 1999). Agenda 21 estimated that the additional investment costs of achieving global water security would be approximately $56 billion per year. The World Bank suggests that $600–800 billion is needed in order to provide basic water supplies and sanitation for all (Cosgrove and Rijsberman, 1999: 78). This certainly constitutes one of the main problems to be addressed in the coming decades. Yet, the main challenge is to reconcile such development needs with environmental pressures. The data in this section will give a clearer idea of the current situation and how difficult it will be to resolve this 'water dilemma'.

Increasing demands are causing water stress in many areas of the world, even in some humid areas, where rising demand or pollution have caused overutilisation of the local resource. According to the Commission on Sustainable Development, water scarcity can be defined as follows:

> Water scarcity occurs when the amount of water withdrawn from lakes, rivers or groundwater is so great that water supplies are no longer adequate to satisfy all human or ecosystem requirements, bringing about increased competition among potential demands. Scarcities are likely to occur sooner in regions where the *per capita* availability of water is low to start with, and with high population growth. They become more serious if demand *per capita* is growing owing to changes in consumption patterns.
> (UNCSD, 1997: § 42)

The Committee on Natural Resources notes with alarm that some eighty countries, constituting more than 40 per cent of the world's population, are already suffering from serious water shortages and that in many cases, the scarcity of water resources has become the limiting factor on economic and social development. According to the Commission on Sustainable Development, about 460 million people, more than 8 per cent of the world's population, live in countries using so much of their water resources that they can be considered to be highly water-stressed. A further quarter of the world's population lives in countries where the use of water is so high that they are likely to move into situations of serious water stress (UNCSD, 1997: § 43). In 1995, water availability was estimated to be 7300 cubic metres per person per year, while in 1970, it was 10,000 cubic meters (UNCSD, 1997: § 34). Figure 2.1 illustrates this alarming trend.

Water scarcity is in fact a recent phenomenon, as shown in Figure 2.1. As stated by the World Water Council, in the 1950s only a handful of countries

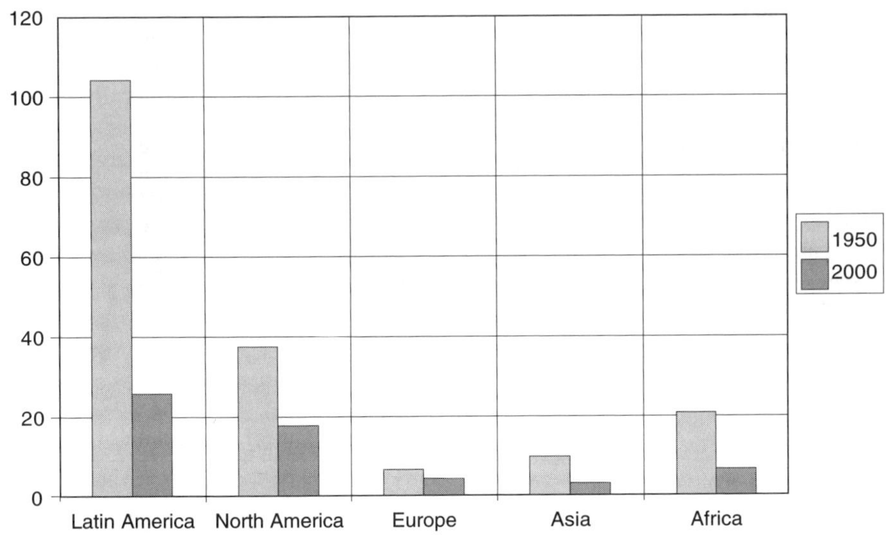

Figure 2.1 Water resources available per inhabitant, 1950 and 2000 (000m^3 per inhabitant). *Source:* FAO (1996: 11).

faced this problem. Nowadays, an estimated 26 countries with a population of more than 300 million suffer from water scarcity. Projections for the year 2050 show that sixty-six countries, comprising about two-thirds of the world's population, will face moderate to severe water scarcity.[12] Table 2.1 shows the current situation regarding water-scarce countries.

We can therefore note an alarming situation, even though the data are sometimes contradictory. Indeed, the seriousness of the data also depends upon the definition of what a 'water crisis' actually is. The figures above referred to what is called 'water scarcity'. Nowadays, however, most water specialists use the concept of 'water stress' rather than water scarcity, since 'water stress' is a broader concept including also qualitative, rather than just simply quantitative, aspects. Box 2.1 defines 'water stress'.

There are also several ways to measure water stress. As described in the World Water Vision Report, the most widely used measurement is the so-called 'Falkenmark indicator', which is the total amount of renewable water

Table 2.1 Countries predicted to have scarce water resources, 2000

		Water availability, m^3 per person	
Country	*Population (million)*	*Water resources*	*Including river flows from other countries*
Egypt	62.4	29	934
Saudi Arabia	21.3	103	103
Libyan Arab Jamahiriya	6.5	108	108
United Arab Emirates	2.0	152	152
Jordan	4.6	153	240
Mauritania	2.6	154	2843
Yemen	16.2	155	155
Tunisia	9.8	384	445
Syrian Arab Republic	17.7	430	2008
Kenya	34.0	436	436
Burundi	7.4	487	487
Algeria	33.1	570	576
Hungary	10.1	591	11,326
Rwanda	10.4	622	622
Botswana	1.6	62	11,187
Malawi	11.8	760	760
Oman	2.3	880	880
Sudan	33.1	905	3923
Morocco	31.8	943	943
Somalia	10.6	1086	1086

Source: FAO (1993).

12 World Water Council: *World Water Challenges for the Twenty-First Century* <http :// www.worldwaterforum.org/Pressreleases/press1.html>.

Box 2.1 Four categories of water stress

- *Low water stress.* Countries that use less than 10 per cent of their available freshwater generally do not experience major stress in respect to the available resources.
- *Moderate water stress.* Use in the range of 10–20 per cent of available water generally indicates that availability is becoming a limiting factor, and significant efforts and investments are needed in order to increase supply and reduce demand.
- *Medium-high water stress.* When water withdrawals are in the range of 20–40 per cent of the water available, management of both supply and demand will be required in order to ensure that the uses remain sustainable. There will be a need to resolve competing human uses, and aquatic ecosystems will require special attention in order to ensure that they have adequate water flows. Developing countries in particular will need major investments to improve water-use efficiency, and the portion of gross national product (GNP) allocated to water resources management can become substantial.
- *High water stress.* Use of more than 40 per cent of available water indicates serious scarcity, and usually an increasing dependence on desalination and use of groundwater faster than it is replenished. This means that there is an urgent need for intensive management of supply and demand. Present use patterns and withdrawals may not be sustainable, and water scarcity can become the limiting factor to economic growth.

Source: UNCSD (1997: § 71).

resources *per capita* per year. Yet, this indicator gives only a rough indication of the current situation. A more precise indicator is the so-called 'Critical Ratio', which is used by the UN in its comprehensive freshwater assessment reports. This indicator measures the ratio of withdrawals for human use in comparison with renewable resources. Both indicators, however, do not correctly reflect each country's water resources management policies. The most precise indicator is thus the 'Current Basin Use Factor', which relates the total consumptive use with the primary water supply. This indicator is particularly useful, as it allows one to separate two different kinds of water scarcities – i.e. on the one hand countries facing physical water scarcity and on the other countries facing economic water scarcity (Cosgrove and Rijsberman, 1999).

Such indicators help us better assess the current situation, and identify more precisely the main challenges to be addressed in the future. In our view, one must distinguish between two such challenges, namely the *development* challenge on the one hand and the *environmental* challenge on the other. On the development side, the growing need for water is increasingly owing to population growth, to urbanisation and to industrial and agricultural development more generally. According to the World Water Council, in 2025 the diversion of water will increase by 18 per cent above the 1995 levels, and consumption by and for humans will increase by 13 per cent (Cosgrove and Rijsberman, 1999: 7). Moreover, according to Sandra Postel *et al.*'s estimations, humans have already appropriated 54 per cent of accessible runoff on earth; future population growth and economic development could lead humans to use more than 70 per cent of accessible runoff by 2025 (Postel *et al.*, 1996: 785). According to the WaterGAP model simulations done for World Water Vision, it is estimated that if water is still managed the same way as today, about 4 billion people, or half of the world's population, will soon be living in countries where more than 40 per cent of the renewable resources are withdrawn for human uses (Cosgrove and Rijsberman, 1999: 41).

Human, and especially industrial, pressure on water resources seems unstoppable. Yet, a better allocation and distribution of water resources can, to a certain extent, resolve this growing problem. But one should also mention here the issue of sanitation, a problem which will also increase parallel to population growth and industrial development. As a matter of fact, the sanitation problem might be even more pressing than the drinking water problem, given in particular the high costs and thus investment needs for sanitation, investments which can come only from the private sector. Overall, one can say that water and health constitute the main challenges along the development paradigm.

On the environmental side, there is a need for a more sustainable way to manage water resources. The current trend indicates that radical action is needed in order to attain this objective. Indeed, Mark Rosengrant, research fellow at the Washington, DC-based International Food Policy Research Institute (IFPRI), concluded in his report, *Water Resources in the Twenty-First Century: Challenges and Implications for Action*, that the number of countries experiencing 'water stress' will double in the next thirty years, and the number of people affected by water scarcity will increase tenfold.[13] Without substantial action, large-scale conflicts and other catastrophes resulting from water shortages may result. As in the case of the development challenge, and as already seen above, the answer to this environmental challenge is treating water as an economic good, the assumption being that economic pricing will lead to a more efficient use of water resources.

In the environmental area, the main challenge is in the agricultural sector. Agriculture is indeed the largest user of freshwater. At present 70 per cent of the total freshwater in the world is used to provide food, natural fibres and employment to billions of rural dwellers (Cosgrove and Rijsberman, 1999: 20). In particular, irrigation consumes a large share of the water withdrawn,

13 *Report finds world water supplies dwindling while demand rises; world food production, health, and environment at risk* (9 March 1997) <http ://www.cgiar.org/ifpri/pressrel/2030997>.

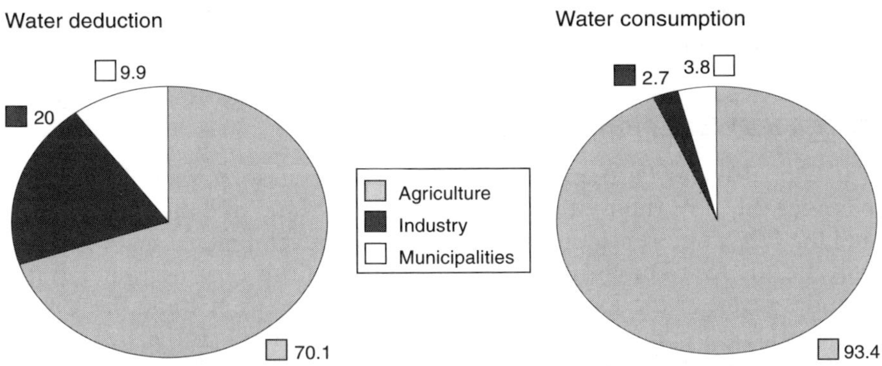

Figure 2.2 Water deduction and water consumption in the world from a sectoral perspective. *Source*: OMM (1997: 8).

both through evaporation from reservoirs, canals and the soil and through incorporation into and transpiration by crops. The water which does not evaporate will recharge the groundwater or contribute otherwise to drainage or return flows. This water can certainly be reused, but it has higher salt concentrations and is often contaminated with nutrients, sediments and chemical contaminants (pesticides, herbicides), which may be even more prejudicial to the ecosystem. Moreover, in their search for suitable lands and adequate water supplies for crop production, millions of farmers have increased the rate of deforestation and desertification, leading to further environmental degradation and further threatening food security. Figure 2.2 illustrates the significant percentage of water used in the agricultural sector.

In conclusion, one may say that the pressures and challenges highlighted above show how quickly action is needed so as to better use the water resources available, as well as to find new forms of co-operation and collective problem-solving. Let us therefore now examine the different actors involved in the water sector, as well as the new dynamics and changes among these actors, which have emerged since the Dublin principles.

A world water vision with a sense of urgency: new organisations and institutions

The current urgency in the water sector has given rise, in recent years, to a series of new organisations and institutions, many of them at the global level. As such, the water sector entirely follows the trend towards global NGOs and TNCs as outlined in Chapter 1. In this section, we will discuss the three main NGOs which have emerged since Dublin, i.e. the Global Water Partnership (GWP), the World Water Council (WWC), and the World Commission on Dams (WCD). In the final section we will then feature a whole set of other organisations also active in the water sector, and so highlight the gradually emerging interpretation of the Dublin Principles.

The Global Water Partnership

The Global Water Partnership (GWP) represents in some ways the achievements of the 1992 International Conference on Water and the Environment and the UN Conference on Environment and Development combined. Not only are the four Dublin Principles embedded in this organisation, but it is also the result of the joint efforts of both the World Bank and the United Nations Development Program (UNDP). The GWP was officially inaugurated in August 1996 at the Stockholm Water Symposium. The idea of such a 'partnership' to address international water problems had in fact already been proposed in August 1995 at the Stockholm Water Symposium organised jointly by the World Bank, UNDP and international water experts. All three agreed to set up a new organisation, which would help co-ordinate the different actors involved in water issues, thus ensuring that the limited financial resources would be used efficiently (see Gleick, 1998: 165–6).

Like the World Water Council (see p. 41), the GWP is relatively new. Consequently, very few documents are available on these organisations. Moreover, both institutions are in the making and no concrete projects or proposed actions seem to have been organised by either, apart from the highly mediated Second World Water Forum in The Hague in 2000.

One particular characteristic of the GWP is the fact that it represents a new type of organisation, which may explain why action is relatively slow: its institutional structure is indeed interesting, as the GWP seeks to regroup all actors involved in the water sector. Consequently, membership of GWP is open to development agencies, international organisations, NGOs, professional associations and academics, as well as the private sector. Quite logically, the GWP's aim is to foster partnerships among all its members. The organisation hopes that such diverse partnerships will enable an integrated water resources management policy and practice as spelled out in the Dublin Principles. Moreover, the organisation rests on a decentralised and on a co-operative structure. As emphasised by the GWP's first executive secretary, Johan Holmberg,

> any entity that respects geographic diversity and cultural priorities cannot rely on a central structure of management or governance, hence the term 'partnership' has been adopted in the truest sense. GWP is building a future wherein the members of the Partnership work with one another, as they deem appropriate, within their own regions and according to the basic parameters as set forth by the Dublin–Rio principles.
>
> (GWP, 1998: iv)

In terms of governance, the GWP defined itself as a 'light, flexible structure, with the emphasis on substance rather than process' (GWP, 1998: 8). The GWP is therefore organised around four different bodies, each having clearly defined roles and responsibilities. The highest decision-making body of the GWP is the Consultative Group (CG), regrouping the representatives of all the actors involved in the water sector. The CG meets annually under the presidency of the GWP chair, who is elected every two years. Decisions are taken to the greatest extent possible by consensus. The first GWP chair was Dr Ismail Serageldin, the Vice-President of the World Bank. He was replaced in August 2000 by Margaret Catley-Carlson, who acted previously as the president of the Canadian International Development Agency (CIDA), as well as Deputy Executive Director of UNICEF and chair of the Water Supply and Sanitation Council. Furthermore, she was also Chair of the Water Resources Management Advisory Committee of Suez Lyonnaise des Eaux.[14] Despite being the highest decision-making body, this Consultative Group is basically a forum for debate, and decisions are taken by consensus. Not surprisingly, the goals and objectives of the GWP remain vague. Consequently, the daily business seems to be run by a Steering Committee, which, *de facto*, acts as an executive body. The third body is the Technical Advisory Committee, which provides professional and scientific advice to the Consultative Group. In particular, the Technical Advisory Committee advises the Consultative Group about priority setting and the identification and performance evaluation of the GWP's Associated Programs. It consists of twelve water professionals and scientists, chaired by Dr Torkil Jonch-Clausen of the VKI Water Quality Institute of Denmark. Since its inception, the Technical Advisory Committee has mainly been operating in Southern Africa and in South Asia, where it has recommended management approaches at the national level, at the river basin level, as well as methods to enhance water conservation and savings. These activities are part of a new co-operation programme taken up by the GWP in collaboration with the Asian Development Bank (ADB) in South East Asia (see Gleick, 1998: 168–70). The Technical Advisory Committee is now working on the African continent with the Water Utilities Partnership for Capacity Building in Africa.[15] The fourth body is the GWP Secretariat, which is based in Stockholm. The Secretariat provides support, administers the GWP governance structure, maintains contacts with organisations and individuals interested in the GWP and organises meetings (see Gleick, 1998: 170).

Why was the GWP created and what are its goals? According to a 1997

14 *Global Water Partnership Names Canadian as New Chairperson.* GWP Press Release <http ://www.gwpforum.org/Hot0822.htm>.

15 *Newsflow*, Global Water Partnership, 2/99: 3.

UNCSD report, the Partnership is an international mechanism aiming at translating the consensus on water management into responsive, coherent services to developing countries, with an emphasis on local implementation (UNCSD, 1997: § 37). Moreover, the GWP will support integrated water resources management programmes by collaborating with governments and existing networks. It will also develop new institutional arrangements and encourage governments, aid agencies and other stakeholders to adopt consistent and mutually complementary policies and programmes, as well as help them share their information and experience. It appears that the GWP has mainly been created in order to accelerate the national and regional co-ordination of water resources management, as well as to encourage these organisations to put the Dublin Principles into practice. More precisely, the following considerations seem to have guided creators of the GWP:

- The growing water crisis in large parts of the developing world;
- the perception that the international response had been fragmented and inadequate, and that this was intolerable in the face of declining financial resources for international development cooperation;
- the recognition that the challenges involved in implementing sustainable water resources management programs in practice were too numerous and too complex to be tackled by any one organization;
- the need for improved coordination among the competing uses of water within and among countries in a situation of increasing demand and growing shortages; and
- the conclusions derived from the Dublin Principles and Agenda 21 that water resources would have to be managed, increasingly, according to an integrated approach, i.e., one that emphasises demand management and negotiations, rather than solutions based on engineering and capital-intensive investments.

(GWP, 1998: 2)

In short, the GWP is an umbrella organisation which regroups the different interested parties and serves as a forum for their discussions. In addition, the GWP also provides services for proposed projects in terms of technical assistance, capacity building, mobilisation of international expertise and funding through its so-called Associated Programs (AP). The GWP also would like to be seen as a sub-regional actor, as the organisation indeed believes that regionalism will be much more important and that water issues can better be addressed at this level. Therefore, the GWP would like to foster a Southern Africa Water Partnership, a South Asian Water Partnership, a South American Water Partnership and so forth (GWP, 1998: 6). The GWP further argues that regional water partnerships would lead to corresponding national water partnerships. However, in our view, the need to create such networks remains unclear. Why not take, for example, the river basin as the management unit, rather than the political units of regions and nation-states, as encouraged by the International Network of Basin Organisations (INBO)? Indeed, integrated water resources management policies at the level of river basins are an essential tool for sustainable development.

In short, this ambitious programme of the GWP reflects a mission and a goal, which is still unclear to many observers.[16] Indeed, the GWP, according to its own words, seeks to:

- Clearly establish the principles of sustainable water resources management;
- identify gaps and stimulate its partners to meet critical needs within their available human and financial resources;
- support action at local, national, regional or river-basin level that follow principles of sustainable water resources management;
- help match needs to available resources; and
- strengthen mechanisms for sharing information and experience.

(GWP, 1998: 3)

One may question whether these lofty goals will ultimately contribute to meeting the pressures and challenges as identified in the second section of this chapter. We must also observe that these objectives are very close to those of a think tank, which constitutes yet another problem, as the objectives of the newly created World Water Council are basically the same.

The World Water Council

The World Water Council (WCC) is usually presented as an international water policy think-tank. This organisation was established as an NGO in 1996, though its origins go back to a meeting in Bari, Italy, in 1995. As a matter of fact, the idea to set up such an organisation was first discussed in 1991 and it was during the UN International Conference on Water and Environment in Dublin in 1992 that the Canadian delegation proposed the formation of the WWC. A Founding Committee, which included a number of water-related associations – such as the International Water Resources Association (IWRA), the International Association on Water Quality (IAWQ), the International Commission on Irrigation and Drainage (ICID), the International Water Association (IWA), but also the World Bank, the Canadian International Development Agency (CIDA), and UNDP – was established at the eighth IWRA Congress in Cairo in 1994. Dr Mahmoud Abu-Zeid, Chairman of the National Water Research Centre in Cairo, and then

16 See for example Ricardo Petrella, 'La nouvelle conquête de l'eau', *Le Monde Diplomatique* (March 2000: 32).

President of the IWRA, was appointed as Chairman of the Interim Founding Committee (WWC, 1996).

According to the WCC, its creation is the result of nearly two decades of a growing global consensus over the need for an international umbrella organisation devoted to the study and the analysis of global long-term water policy issues and the advocacy of better management of world water resources (WWC, 1997). From its origin, the WCC gave itself the following seven objectives, which were subsequently written out in its draft constitution:

- To identify critical water issues of local, regional and global importance on the basis of ongoing assessments of the state of water in all its dimensions.
- To promote sustainable mechanisms to raise awareness about critical water issues at all levels of decisions making, including the general public in different parts of the world.
- To facilitate arrangements for providing advice and relevant information to institutions and decision-makers concerning actions, which are required to ensure the sustainability of global, regional and national water resources.
- To sponsor programmes and actions to promote policies and strategies for conservation of water as well as comprehensive and integrated water resources planning and management, and to contribute to the resolution of issues related to the transboundary waters, including river and lakes basins, aquifers and wetlands.
- To assist in ensuring availability and accessibility of water of suitable quality for the poor and disadvantaged population and ensure social and gender equity in respect to water use.
- To promote the implementation of effective policies and strategies for integrated management of water demand, and encourage appropriate actions by concerted organisations.
- To provide a forum to arrive at a common strategic vision and platform to launch actions required for sustainable management of water resources worldwide (WWC, 1996).

So far, the main activities of the WCC have been to organise the so-called World Water Forum, which has been held twice since. The main objectives of this Forum are (1) to raise awareness among decision-makers, the media and the public at large of world water issues of critical concern, (2) to provide a platform for exchanging views, information and knowledge on current issues related to global water, (3) to extend the opportunity to the Council to promote its policies and findings and to share with other organisations a common vision and interests in world water affairs and (4) to present state of the art knowledge on global water assessment, challenges and potential solutions.[17]

The first such World Water Forum was held in Marrakesh in March 1997. The Forum, entitled 'Water: World's Common Heritage and Future' was guided by two main objectives, namely to raise international awareness at all levels, but in particular among world leaders and decision-makers on global freshwater issues, and to create a long-term vision for water resources management (World Water Council, 1997: 3). More concretely, the First World Water Forum launched, one year later, a so-called 'Vision exercise' to lead to a report entitled 'Vision for Water, Life and the Environment'. A 'World Commission on Water in the 21st Century' was set up to accompany and guide the exercise. Again, the main goals of this vision exercise pertain to awareness raising, policy advice for decision-makers and funding agencies, and of course to producing a 'Vision for the year 2025 that is shared by water sector specialists and decision makers in the government, the private sector, and civil society' (Cosgrove and Rijsberman, 1999: Annex A). The process began with four thematic panels in which experts discussed possible developments affecting the water sector in the future. They were convened around the issues of (1) Energy Technology, (2) Information and Communications Technology, (3) Biotechnology and (4) Institutions, Society and the Economy. The output from the thematic panels were integrated into possible scenarios at the global level. The results of these discussions was then synthesised into a draft Vision, which was discussed and criticised[18] during the 1999 Stockholm Water Symposium,[19] before it was finalised and presented at the Second World Water Forum and Ministerial Conference.

The Second World Water Forum was organised in the year 2000 in The Hague. During the Forum, the Dutch government also organised a Ministerial Conference on water management in the twenty-first century, entitled 'From Vision to Action'. The Ministerial Conference was thus set up to provide the political stage for a debate about the above mentioned vision exercise. This Ministerial Conference, not surprisingly, ended with a Ministerial Declaration on Water Security in the 21st Century, which mainly advocates

17 <http ://www.worldwaterforum.org/Pressreleases/press9.html>.

18 Indeed, according to the Final Conclusions of the 9th Stockholm Water Symposium by the Stockholm International Water Institute (SIWI), the World Water Vision was at this time largely unrepresentative of the world population since, in the own words of SIWI, 'Since the Vision in its present form represents the efforts of a large community of senior men, it has a long way to go in its second round of consultations to include also the remaining 50% of the world population in the consensus. This predicament shows that the world is still rather far away from the feminization of water management and policies, which was recommended at the First World Water Forum in Marrakesh' (Final Overall Conclusions, The 9th Stockholm Water Symposium, 9–12 August 1999, Stockholm International Water Institute website, <http://www.siwi.org>).

19 The Stockholm Water Symposium is one of the most important annual meetings on global water issues. The Symposium is organised by the Stockholm International Water Institute, which in turn is linked to the Stockholm Water Company.

integrated water resources management, as well as stresses the need for further co-operation on water issues within the UN system.[20]

But the Second World Water Forum was also the opportunity to unveil the 'Vision' to the public. On the one hand, the public learned, this vision exercise was highly successful, inasmuch as it managed to involve more or less all relevant 'partners' in the water sector. On the other hand, however, the results of this exercise are as lofty and fuzzy as is the The Hague Ministerial Declaration. Indeed, as a result of this exercise, the World Commission on Water in the 21st Century has managed to identify the following seven core challenges in the global water sector: (1) water scarcity, (2) lack of accessibility, (3) water quality deterioration, (4) world peace and security, (5) awareness by decision-makers and the public, (6) decline of financial resources allocation and (7) fragmentation of water management.[21] Based on this truly impressive achievement, the World Water Council announced a new initiative for putting the World Water Vision into practice, among which were the following three suggestions:[22]

- To establish a monitoring system for the World Water Vision in Action. To this effect a World Water Vision Monitoring Secretariat will be set up.
- To stimulate policies for the financing of water development and protection. More concretely, one can read:

  > The World Water Council proposes to unleash the power of the private sector, focusing on an 'enabling environment', with a regulatory, legal and institutional framework that will promote and protect investments . . . Money is available but to attract investors a predictable, transparent regulatory framework and reasonable returns on investments must be provided. The establishment of a global risk management fund will further help to protect potential investors.[23]

- To create a World Commission on Water, Peace and Security.

One can thus note with astonishment that water protection, investor protection and world peace all go hand in hand when put under the auspices of the World Water Council. It is of course true that the Water Vision Exercise

20 For the whole declaration, see 'Declaration of The Hague' <http ://worldwaterforum.org/Ministerial/declaration.html>.

21 Second World Water Forum, 'World Water challenges for the twenty-first century', World Water Forum Press Release 1 <http://www.worldwaterforum.org/Pressreleases/Press1.html>.

22 World Water Council, 'The World Water Council announces new initiatives for putting the world water vision into action' <http ://www.worldwaterforum.org/Pressreleases/press2.html>.

23 Ibid.

was highly ambitious, and thus by definition quite unspecific. And to a certain extent, this exercise can even be labelled a success. As the Prince of Orange, Chairman of the World Water Forum and Ministerial Conference remarked, in his concluding note at the Second World Water Forum in The Hague: 'What began as an ambitious idea has become a tremendous success . . . Never before, has the world seen such a large gathering of water specialists, concerned citizens, politicians and organisation representatives. In total, more than 5000 participants were counted in The Hague, a figure that far exceeded our best expectations'.[24]

On the other hand, groups such as Public Services International made the observation that the World Water Forum, with its exorbitant entrance fees, was 'not everybody's business'.[25] One may also note with astonishment that the 10th World Congress of the International Water Resources Association (IWRA) was held in Melbourne, Australia at exactly the same time. One explanation may be that there could be a divide here between on the one hand the water specialists of IWRA who are particularly concerned about the South and on the other hand the WWC which, as we have seen, seems to be particularly interested by investor protection, world peace and media coverage. Indeed, one could also criticise the fact that the Ministerial Declaration remains a declaration with no legal value.

In short, we can see that in fact the WCC's main activity is more or less limited to organising the different World Water Fora, and thus to awareness-raising on problems related to water resources management. Yet, the WWC is also launching two ambitious long-term projects: the first is the 2025 Vision for Water, Life and the Environment, a process that will, according to the WCC, lead to the systematic identification of water-related problems facing the world in the twenty-first century and corresponding recommendations. The second main project of the WWC is to create a Global Water Assessment that would, again according to the WCC, develop for the first time a satisfactory inventory of the world's freshwater resources. Besides these two main projects, the WCC will mainly be a public relations exercise by means of conferences, as well as by means of publishing its journal (entitled *Water Policy*), as well as a *Yearbook*.

Both the WWC and the GWP were set up after Dublin, in order to create worldwide awareness about the looming 'water crisis'. Both have certainly contributed to such international awareness-raising, as well as doing extensive monitoring work. However, a little more than five years after their creation, one may still question their legitimacy and usefulness. Indeed, as a

24 World Water Council, 'The Forum Reviewed' <http ://www.worldwaterforum.org/Reviewed/index.html>.

25 For more criticism, see 'Controlling the Vision and Fixing the Forum: the politburo of privatisation', World Water Forum, The Hague (17–22 March 2000), PSI Briefing: Critique of the Vision <http ://www.psiru.org>.

first criticism, one may ask what actually distinguishes them. The following is the GWP's own answer to this question:

> In the minds of some, there is confusion as to the roles of the GWP and the [World Water] Council. Why is there a need for both? Are they not competing for attention in the same arena? The answer is that both are needed. If one does not exist, the other would quickly call for its creation. They did not compete, but support each other mutually in the same arena. Both have the objective of promoting sustainable management of the world's water resources, but they are serving it in a complementary manner.[26]

In turn, the WWC also felt compelled to reply in its first newsletter, declaring that both had a distinctive role, one (the WWC) being a think-tank on a long-term basis, while the other (the GWP) translating this 'thinking' into field-based programmes to developing countries.[27] While both indeed, appear to have different objectives, the WCC seems to have no clear vision, while the GWP seems to have no clear action plan and projects. Indeed, after five years of existence, both websites are still under construction and, apart from a four-page newsletter, no information is given on the work agenda or on the different ongoing projects. Some scepticism about the exact role of these two organisations is thus in order.

The World Commission on Dams

The World Commission on Dams (WCD) differs slightly from the two other organisations in the sense that it was created for just a two-year period. It is thus a short-term project. The final report of the Commission was released on 16 November 2000, and the Commission was actually decommissioned at the launch of the report. However, like the GWP and the WWC, the World Commission on Dams has been built around the idea of a consensus to be reached among all the interested parties. Such a consensus has in fact been made necessary, as dam building has become a highly disputed issue and has led to clashes between dam proponents and critics. Indeed, there has been mounting controversy since the 1980s about the role of large dams in development. As one of the critics says:

> Where dams have been built to generate electricity, the justification has been that they provide a cheap, reliable and economic source of power – a

26 *Newsflow*, Global Water Partnership, 1.
27 World Water Council, *Newsletter*, No. 0 (March 1997: 5).

> claim long disputed by critics of large dams. Such critics point out that slippage in construction schedules and over-optimistic assessments of output have made many dams costly white elephants, saddling Southern countries in particular with unmanageable debts which poorer people invariably end up paying for, through cuts in public expenditure on health, education and other public services.[28]

The main target of this critique is in fact the World Bank, as 'The World Bank has supported around 10 percent of large dams in developing countries, and it has a leading role in funding some of the largest facilities anywhere. The Bank has directly funded four out of the five highest dams in developing countries outside China, three out of the five largest reservoirs in these countries, and three of the five largest hydroplants' (Gleick, 1998: 175–6). It is in this context that the World Bank decided in 1995 that its own Operations Evaluation Departments (OED) should undertake a review of large dam projects. In September 1996, the World Bank released the review of fifty World Bank-funded large dams projects. In the Bank's own review, although most of the projects met the standards applicable when they were approved, only 25 per cent were implemented in a way that complied with current World Bank policies.[29] Upon the initiative of David McDowell, at that time Director General of IUCN, the World Conservation Union, a general partnership between IUCN and the World Bank was set up: on the basis of this OED review, both organisations decided to set up a workshop regrouping thirty-nine participants including government representatives, the private sector, international financial institutions, and civil society.[30] One of the most important recommendation of the workshop was to establish a two-year mandate for a World Commission on Dams (WCD) which would (1) review the development effectiveness of dams and assess alternatives for water resources and energy development, and (2) develop internationally-accepted standards, guidelines and criteria for decision-making in the planning, design, construction, monitoring operation and decommissioning of dams.[31]

As an initial step, the IUCN and the World Bank established an Interim Working Group and appointed Minister and Professor Kader Asmal from

28 CornerHouse Briefing, *Dams On the Rocks: The Flawed Economics of Large Hydroelectric Dams* (8 August 1998). See also Goldsmith and Hildyard (1986); McCully (1996, 1997a, 1997b).

29 Liebenthal *et al.* (1996).

30 *About us: History-Workshop Summary.* World Commission on Dams <http ://www.dams.org/summary.asp>.

31 *The Mandate of the WCD: Goals and Objectives.* The World Commission on Dams <http ://www.dams.org/mandate.asp>.

South Africa as future chairman of the Commission. Indeed, the very creation of the Commission was controversial, as the World Bank document evaluating its own dam building projects was highly criticised. For example, the International Rivers Network issued a statement, endorsed by forty-nine other NGOs from twenty-one countries, putting in question the methods and data used in the OED report.[32] Moreover, the names of the ten potential commission members proposed in October 1997 drew immediate opposition, especially from some of the NGOs who felt that the proposed list had 'inadequate representation of dam-affected people and anti-dam NGOs' (Gleick, 1998: 179).[33] Finally, on 16 February 1998, following the fourth meeting of the IUCN/World Bank Interim Working Group, the WCD was formally launched in Cape Town, South Africa.

The WCD is composed of a secretariat and 12 commissioners.[34] The Commission's main work will be to examine 150 large dams selected for case studies in terms of their environmental, economic, and social effects. More precisely,

> The Commission will be independent and its remit will include issues that address both broader considerations such as water and energy policy as well as more specific technical and case study oriented questions, such as resettlements, compensation of affected communities, ecological impacts, and wider effects in basin-wide catchment contexts. The Commission's work will be of an advisory nature, not investigatory in the sense of judicial commissions.[35]

However, even though there is a strong consensus for an independent assessment of dams and dam policy, the issue has already become highly politicised. Not surprisingly, the main conclusions of the report are disappointing: the report highlights three basic facts, i.e. the fact that we now have the first comprehensive global and independent review, the fact that the WCD has shifted the debate to the performance and legacies of existing dams and the fact that the importance of a participatory decision-making approach using a right-and-risks approach is now recognised (WCD, 2000). In other words, the WCD has simply legitimised the construction of large dams, as long as they recognise and measure the social and environmental effects. The

32 *The World Commission on Dams: Summary. What Happened?* The World Commission on Dams <http ://www.dams.org/summary.asp>.

33 For more details, see International Rivers Network: *NGOs Condemn World Bank/IUCN Dam Building Commission*, Press Release (25 November 1997), Berkeley, California.

34 See Appendix 4 (p. 240).

35 *The Mandate of the WCD: Scope of Work*. The World Commission on Dams <http :// www.dams.org/mandate.asp>.

development of alternatives to dams and electricity consumption have either been ignored or dismissed.[36]

All three new organisations in the water sector – the WWC, the GWP, and the WCD – share the same idea, namely that all concerned actors and stakeholders need to be included as partners, both in diagnosing and in making recommendations. In doing so, they all follow the logic of the Dublin Principles. But like the Dublin Principles the mandates of all three organisations are sufficiently vague, so as to allow a general consensus, yet leaving the operational content undiscussed and probably unrealisable. Also, as the list of names in the Appendices show, these three bodies often have the same people as members, or in any case a very similar type of people, former politicians, affiliates of international environmental NGOs and other international dignitaries. One notes, however, an under-representation of the water specialists with a natural science or an engineering background. We thus would like to analyse further still other actors of the water sector, who, in our view, are probably more important than the three institutions already discussed.

The broader international water arena

In this section we want to look at other major actors in the international governmental water arena, except for the World Bank, which will be dealt with in Chapter 3. Of course, it is impossible to present all the major actors. We will thus focus on the ones belonging to the larger UN system, plus a few selected others. Within this UN system, the main inter-governmental bodies providing policy and guidance in the field of water resources are the General Assembly and the Economic and Social Council (ECOSOC) and two of its subsidiary bodies, the Commission on Sustainable Development and the Commission on Natural Resources, as well as the regional commissions (UN ECOSOC, 1997: § 9). Specifically dealing with freshwater is an Administrative Committee on Coordination (ACC) Sub-Committee on Water Resources. This sub-committee is composed of all organisations and bodies of the United Nations system involved in the field of freshwater resources. In regard to the modalities for the preparation of the reviews of Chapter 18 of Agenda 21 – i.e. the Agenda 21 chapter which deals with water – the ACC Sub-Committee, at its fourteenth session, agreed that the UN Secretariat would act as a focal point for the area of integrated water resources development and management, i.e. the United Nations Educational, Scientific and Cultural Organisation (UNESCO) and the World Meteorological

36 'Obstacles to the adoption of these options range from market barriers to institutional, intellectual and financial barriers. A range of incentives – some hidden – that favour conventional options limit the adoption rate of alternatives' (WCD, 2000: xxxiii).

Organization (WMO) for water resources assessment, the World Health Organization (WHO) for the protection of water resources, water quality and aquatic ecosystems, as well as for drinking-water supply and sanitation, the United Nations Center for Human Settlement (Habitat) for water and sustainable urban development, the UN Food and Agriculture Organisation (FAO) for water for sustainable food production and rural development and the WMO for the impact of climate change on water resources. In regard to cross-cutting issues, it was agreed that the International Research and Training Center for the Advancement of Women (INSTRAW), the United Nations Children's Fund (UNICEF) and UNESCO would act as focal points for issues concerning women's communication, education and research, respectively (UN ECOSOC, 1997: § 16). It can be clearly seen from this list that the UN are basically approaching water from a scientific, a health, and a development approach. We will discuss the respective UN organisations in this order. However, as seen earlier in this chapter, the main concern, which has triggered water issues globally is the environment. We will discuss this environmental approach within the UN at the end of this section.

Science

Historically, the scientific approach has always been dominant at the international level. Many of the actors here are academic professional associations or universities. Within the UN system, the main actors with a scientific approach are UNESCO, as well as the WMO. More recently, UNESCO and WHO have also started to collaborate on certain scientific water projects. In conclusion, we will also briefly mention the International Water Resources Association.

UNESCO

Since its very origin in 1945, UNESCO has focused on scientific questions. Consequently, since the late 1950s, UNESCO has dealt with the scientific side of water resources management, in particular through its Intergovernmental Oceanographic Commission (IOC), the Coastal Regions and Small Islands Project, and the International Hydrological Programme (IHP), which is the one we are most interested in here. The IHP was created in 1975 to follow up on the International Hydrological Decade (1965–75) with the purpose of bringing the international scientific community and the Member States together in an effort to promote 'the development of methods and the human resource base for the rational management of water resources, including the protection of the environment'.[37] From the 1990s on, we can observe a shift in

37 *UNESCO, About IHP: Mandate and Scope of the International Hydrological Programme* <http ://www.pangea.org/orgs/unesco/newaboutihp.html>.

UNESCO's (just as in all UN) policies, from development to 'sustainable development': 'greater emphasis is [now] being put on the role of water resources management for sustainable development and the adaptation of the hydrological sciences to cope with the expected changing climate and environmental conditions'.[38] The IHP is currently in its fifth programme (IHP–V). In this programme, UNESCO wishes to place more emphasis on water resources management, rather than focusing on hydrology and its related factors as previously. From the IHP–V outline, integrated water resources management appears as one of the four major issues, although it not exactly clear how the IHP can contribute to this far broader issue given the quite narrow scientific perspective followed until now. In short, UNESCO appears to have substantially broadened its mandate.

WMO

Alongside UNESCO, another UN organisation has taken on a growing role in this scientific approach to water. Indeed, water now undisputedly constitutes one of the main missions of the WMO agenda. All WMO's activities in the water sector are currently regrouped under the Hydrology and Water Resources Programme (HWRP). Indeed, as stated by the WMO, the 'Hydrology and Water Resources Programme fulfils one of the major purposes of WMO, namely "to promote activities in operational hydrology and to further close cooperation between Meteorological and Hydrological Services"'.[39] More precisely, the

> activities under the HWRP concentrate on the measurements of basic hydrological elements from network of hydrological and meteorological stations; the collection, processing, storage, retrieval and publication of hydrological data, including data on quantity and quality of both surface water and groundwater; the provision of such a data and related information for use in planning and operating water resources projects; and the installation and operation of hydrological forecasting systems.[40]

In short, the WMO increasingly seems to focus on identifying the world's freshwater resources, and develop corresponding simulation and forecasting mechanisms.

WMO also works with the FAO on such topics as data processing and with the International Atomic Energy Agency (IAEA) on safety standards for power plants, response to nuclear accidents and the application of isotopes in

38 Ibid.

39 Paragraph (e) of Article 2 of the Organisation's Convention.

40 *Hydrology and Water Resources Programme (HWRP) as approved by the Thirteen WMO Congress (Cg–XIII) in May 1999* <http ://www.wmo.ch/web/homs/hwrpdesc.html>.

operational hydrology.[41] The WMO has also agreements with professional associations, institutes such as the International Association of Hydrological Sciences (IAHS), the International Organisation for Standardisation (ISO), the International Institute for Applied Systems Analysis (IIASA), the International Association for Hydraulic Research (IAHR), the International Commission on Irrigation and Drainage (ICID), the WWC, the GWP and international river basin commissions such as INBO.[42] These new collaborative programmes reflect the increasing need to gather information at the international level.

The most important monitoring activity has in fact been taken up by UNESCO and WHO, which is starting also to get involved within this scientific approach. In this effort, a Joint Monitoring Programme for Water Supply and Sanitation (JMP) was established through a protocol signed between the WHO and UNICEF in 1990.[43] The main goal of this programme is to build information systems at the country level on water resources. This programme is of considerable importance since recent studies by the UN have shown that many national hydrological services and agencies, particularly but not exclusively in developing countries, are faced with severe financial and staffing reductions, leading to poor monitoring activities.[44] Since 1991, the countries involved are now integrated into a computer system, the so-called Water and Sanitation Monitoring System (WASAMS), which keeps track of

41 *HWRP – Programme on Water Related Issues: Programme Organisation* <http :// www.wmo.ch/web/homs/watrel.html>.

42 International river basin commissions, as promoted by INBO, are a relatively new phenomenon. The Danube river basin, covering seventeen countries, is protected by the Danube Protection Convention signed in 1994 but has no real river basin commission. The most interesting example is the Mekong River Commission created in 1995. It is composed of three permanent bodies. Its mandate is as follows:

- To co-operate and promote in a constructive and mutually beneficial manner sustainable development, utilisation, conservation and management of the Mekong River Basin
- To protect the environment, natural resources, aquatic life and conditions, and ecological balance of the Mekong River Basin from pollution or other harmful effects resulting from any development plans and uses of water and related resources in the Basin
- To define as the 'general planning tool and process' that the Joint Committee would use as a blueprint to identify, categorise and prioritise the projects and programmes to seek assistance for and to implement the plan at the basin level.

These commissions are perhaps the most effective instruments in environmental protection and monitoring (for more details, see Finger and Tamiotti, 2001).

43 *Protection of the Human Environment: WHO/UNICEF Joint Monitoring Programme for Water Supply and Sanitation* <http://www.who.int/water_sanitation_health/wss/ Monitoring3.html>.

44 *Coordination of the policies and activities of the specialised agencies and other bodies of the United Nations: Freshwater, including clean and safe water supply and sanitation.* Report of the Secretary-General, United Nations Economic and Social Council, 9 June 1997, E/1997/ 70, § 19.

the progress accomplished by the different countries. Moreover, the JMP has conducted a research project called the Global Water Supply and Sanitation Assessment 2000 giving the situation in water and sanitation worldwide at the end of 1999.[45] From 2001, this database will be periodically updated. In short, one can see that the general trend among UN scientific actors is to both assess the state of the water resources and to build national capacities to do so.

Finally, one should also mention, among the scientific actors, the International Water Resources Association (IWRA). IWRA is 'a non-profit, non-governmental, educational organisation, connecting professionals, students, and individuals who are concerned with the sustainable use of our water resources'. It furthermore 'provides a global forum for bridging disciplines and geographies'.[46] This association was created in 1972 and is recognised as one of the very important players in the global water sector. IWRA, as we have seen above, was one of the founders of the WWC. IWRA is mainly research-oriented, and as such focuses on education programmes, research, conferences and publication (i.e. *Water International*). According to IWRA itself, it is deeply committed to the sound management of water resources by means of

> providing an international forum for water resource issues, advancing water resources and related environmental research, promoting water education throughout society by improving global access to relevant data and information, enhancing the quality of knowledge used in decision-making, improving exchanges of information and expertise, and networking with other organisations to advance common goals.[47]

Within IWRA's larger scientific network, there are also a number of academic institutions, which play a considerable role in the water sector. We should mention here in particular the Water, Engineering and Development Center (WEDC) situated at Loughborough University (UK) and the International Institute for Infrastructural, Hydraulic and Environmental Engineering (IHE) at the University of Delft in the Netherlands. All these private academic actors generally come from an engineering background, yet have been forced, over recent years, to open up their approaches to other, more societal, aspects. Some of them, like WEDC, have become very good at that.

All these scientific actors do play a key role in the water sector, which must not be underestimated. But only UNESCO, it seems to us, is really willing to

45 Report available on <http://www.who.int/water_sanitation_health/Globassessment/Global/TOC.htm>.

46 International Water Resources Association leaflet.

47 Ibid.

take on a comprehensive role in framing a global water resources management policy from a more scientific perspective. However, given the fact that UNESCO has serious financial problems and is already now spread too thin, one may have doubts as to this organisation's ability to really deliver. Yet, unfortunately, we do not see any other global organisation being able to take on this badly needed function.

Health

Apart from a scientific, a developmental and an environmental approach, water can also be approached from a health perspective. And indeed, water, sanitation, and health does constitute one of the biggest issues in sustainable development. Outside the UN organisations, this concern is mainly being addressed by NGOs, as well as by national organisations of development co-operation. Within the UN, however, health is clearly the responsibility of the WHO. Consequently, the WHO is involved in drinking-water quality, in sustainable water supply and sanitation, in hygiene promotion, in water resources protection, in healthcare waste, in recreational waters and much else besides.[48] More precisely, the WHO is implicated in a substantial number of programmes such as the sustainability and optimisation of water supply and sanitation services, as well as in water supply and sanitation monitoring. Actually, the WHO has been at the forefront of initiatives and programmes dealing with these issue since the 1977 United Nations Water Conference, and in particular during the International Drinking and Sanitation Decade (1981–90), where the organisations prepared two significant reports. However, in light of the realignment of international organisations and institutions, as highlighted in Chapter 1, health increasingly seems to be moving, as does the WHO as an organisation, away from a developmental concern and in the direction of humanitarian and security issues. This means that health concerns in the water sector will more and more be addressed in a crisis mode, rather than in a long-term perspective, parallel to sanitation becoming increasingly important.

Development

Water, however, in essence was and still is a developmental issue, despite the fact that its recent attention stems from environmental concerns. Historically, development, as we have seen in Chapter 1, falls mainly into the domain of the UN. However, UN development agencies have to redefine themselves in light of the growing role of the World Bank in the area of development. As

48 *Protection of the human environment, Water, Sanitation and Health* <http ://www.who.int/water_sanitation_health/index.html>.

such, the evolution of UNICEF and the FAO – here, in the case of water – are quite typical.

UNICEF

UNICEF has been involved in water, environment and sanitation issues since the 1960s. Its first major interventions in this area go back to its response to a devastating drought in northern India in 1967. Since then, UNICEF has supported long-term programmes in some ninety countries in Asia, Africa and the Americas.[49] The main focus of UNICEF is of course on children's rights and articulated around the Convention on the Rights of the Child. UNICEF also focuses on Principle 4 of the Dublin conference (i.e. the full participation of women in their various roles as care-givers, workers, beneficiaries of services and decision-makers in homes, communities and at national levels).[50] At first, UNICEF supported government programmes for the provision of a minimum level of water supply and sanitation for those most in need, which involved primarily the rapid drilling and installation of bore-holes with hand pumps in rural areas. Later, the programmes diversified, during the late 1960s and 1970s, into relatively large-scale national water programmes, including the provision of sophisticated drilling rigs and equipment, gravity-fed systems, protected springs and wells and upgrading of traditional water sources in rural areas (UNICEF, 1995: 7). More recently, UNICEF has refocused its activities in the water sector along the humanitarian and security paradigm, i.e. articulating its interventions around issues of water security, safe environmental sanitation and adequate care and hygiene practices (UNICEF, 1995: 19). Not surprisingly – i.e. because it thinks from a humanitarian aid and security, rather from a developmental perspective – UNICEF is quite critical about the recent trends in water resources management, such as privatisation, when it says that

> while privatisation may bring many benefits in terms of incentives for efficient service delivery, it is not necessarily a comprehensive solution and risks further marginalising the poor. First, there are few incentives to reach out to informal settlements in peri-urban and rural areas. Low incomes and limited 'voice' mean that the poor can bring little pressure to bear. Second, privatisation, particularly in monopoly services, has brought to the fore the need for the regulation of price and service quality, with special consideration to the prices and service level for the poor.
>
> (UNICEF, 1995: 6–7)

49 UNICEF, *Water, environment and sanitation (WES): About WES* <http ://www.unicef.org/programme/wes/info/about.htm>.

50 Ibid.

FAO

A similar pressure to shift from a developmental to a humanitarian and security perspective, again in the water sector, can be found in the case of FAO. Indeed, FAO's historical focus of action was on 'sustainable' food production and rural development. In the water sector, FAO had established the International Action Programme on Water and Sustainable Agricultural Development. This programme was and is still being used as the major instrument for assisting states in implementing the sustainable agriculture programme area of Agenda 21's Chapter 18. More precisely, this Programme focuses on (1) reviewing the national water sector in general and the irrigation sector in particular, (2) identifying issues and opportunities for development in the context of policies/strategies, river basin master planning, adaptation of technologies, management issues, environmental sustainability and national capacity-building and (3) formulating national action programmes to address issues and realise opportunities. In this context, assistance has been provided to formulate national and sub-regional action programmes in China, Egypt, Indonesia, Syria, the United Republic of Tanzania, Turkey, Zimbabwe and the Lake Chad basin. Assistance for initial implementation of the programme has been provided to four of those countries, namely Egypt, Indonesia, Turkey and Zimbabwe (UN ECOSOC, 1997: § 27). However, within the realignment among international organisations and institutions, we can see, and this is particularly the case for FAO, how environmental concerns are redefining the development agenda, thus bringing in the World Bank as a new and increasingly central actor in water sector.

NGOs

Apart from the World Bank, which will be presented and analysed in Chapter 3, water in the development arena has been dealt with, and continues to be addressed by, numerous NGOs, many of them supported by national agencies of development co-operation. The role played by these development agencies should not be underestimated especially the British Development Agency (DFID),[51] the Swedish Development Agency (SIDA)[52] and the Danish Development Agency (DANIDA).[53] SIDA, for example, makes water one of its foremost priorities with a special workforce, 'Sida Water' regrouping the Natural Resources and the Environment Water and Sanitation Division, The Infrastructure and Economic Co-operation Division, the Co-operation with NGOs and Humanitarian Assistance Division as well as the Research

51 See <http://www.dfid.gov.uk>.
52 See <http://www.sida.se>.
53 See <http://www.um.dk>.

Co-operation Division. DFID is certainly the most active, at least on a media level, among these development agencies in the water sector and is especially supporting public–private partnership programmes (DFID, 1997). But the Dutch, the Swiss, the French, and the Finnish development agencies can also be found in the water sector. Yet, as we will show in Chapter 5, many of these NGOs are increasingly becoming sub-contractors to World Bank projects.

Environment

Environment, as we have already seen in this chapter, is the main trigger of the renewed concern for global water issues. Environment is also the means by which – within the context of the World Bank's approach to development – water is being redefined in terms of a managerial issue, hence the new term of 'water resources management'. In this section, we would like to illustrate this evolution by presenting three more actors in the global water arena, namely, UNEP, UNDP and the Global Environmental Facility.

UNEP and UNCHS

The most obvious organisation to address global water issues in the context of Dublin and Rio would without doubt have been the United Nations Environment Programme (UNEP). But then, the Rio process was precisely set into motion to weaken UNEP (see Chatterjee and Finger, 1994), and moreover UNEP was merged with Habitat (UNCHS), an organisation with a totally different mandate – which also happened to be located in Nairobi. Not surprisingly, UNEP's/UNCHS' role in global water (resources management) is at best unclear and vague. Indeed, not much can be found in their programme of action, which makes one wonder why an organisation, which should be at the forefront of international environmental problems, does not play a more important role in the global water arena. These two organisations' most important water initiative is actually the Water for African Cities Programme, an initiative, launched in 1999, and which results from the dialectal relationship between environment and urbanisation. Its main objectives are '(1) to build capacity in seven demonstration cities in the water sector, (2) to avail information on and best practices in urban water management, and (3) to link sector professionals working in the field of water management with each other and with other networks, institutions, governments, municipalities, NGOs and the private sector'.[54] This programme has therefore commenced in seven demonstration cities, i.e. Abidjan (Côte d'Ivoire), Accra (Ghana), Addis Ababa (Ethiopia), Dakar (Senegal), Johannesburg (South Africa), Lusaka (Zambia) and Nairobi

54 *Managing Water for African Cities* <http ://www.un-urbanwater.net/>.

(Kenya),[55] where it is 'improving the efficiency of water use through the introduction of a range of demand management measures, mitigating the impact of urban activities on freshwater resources'.[56] Though this programme contributes to reaffirming the role of UNEP in the water sector, it will not be sufficient to avoid UNEP's overall decline and instrumentalisation by the World Bank, be it in water or other sectors.

WSP

In the same perspective of World Bank dominance in the water sector, we would like to mention the UNDP/World Bank Sanitation Programme (WSP). This programme was created in 1978, and is currently active in more than thirty countries throughout Latin America, Africa, South Asia and East Asia. The WSP began as a set of separate projects supported by the UNDP and implemented by the World Bank to test and promote low-cost appropriate technologies. Its mission statement reads as follows: 'the Water and Sanitation Program helps poor people gain sustained access to improved water and sanitation services. The Program works with partners in the field of innovative solutions to the obstacles faced by poor communities and strives to be a valued source of advice to achieve widespread adoption of these solutions' (UNDP/World Bank, 1999b: 1). One can still clearly detect here UNDP's handwriting. As of 1992, the WSP followed what we can call the 'Dublin spirit', inasmuch as it started treating water as an economic good, increased user participation (which included both community involvement but also private sector participation), and focused on the role of women. Interestingly, the restructuring of the international organisation, and in particular the restructuring of UNDP, led to the end of this UNDP/World Bank collaboration. As of 2000, WSP is now under World Bank tutelage.

GEF

After having mentioned UNEP and UNDP and its collaboration with the World Bank, it is now logical to also mention their joint activities through the so-called Global Environmental Facility (GEF).[57] The GEF was created in

55 *Water for African Cities*, quarterly newsletter, Issue One, UNCHS/UNEP, 'Programme Starts in Seven Cities' (January 2000: 1).

56 Ibid: 4.

57 For more information, see <http://www.gefweb.org>. See also: El-Ashry, Mohamed, *Water Resources: a Global Environmental Perspective*, Chief Executive officer and Chairman Global Environmental Facility, paper presented at the Stockholm Water Symposium, Stockholm, Sweden (16 August 1995); El-Ashry, Mohamed, *Global Environment and Sustainable Development*, paper presented at the American University, Centre for the Study of the Global South, Summer Institute United Nations, New York.

1994 after a four-year pilot phase in the context of Rio. As such, the GEF is the newest addition to the Bretton Woods family of international financial institutions (IFIs).[58] It regroups UNDP and World Bank member countries, and uses the World Bank, UNEP and UNDP as the implementing agencies. More precisely, the GEF is involved in four focal areas, i.e. climate change, biodiversity, ozone depletion and pollution of international waters, all regrouped under the term of 'sustainable development'. Within this regrouping, it is however more than obvious, as we will see in more detail in Chapter 3, that it is the World Bank, which plays the preponderant role among the three, and this is particularly the case in the water sector. In reality, the GEF might just as well be a new multilateral aid agency for the World Bank to meet the governments' promises at the Rio Earth Summit.

NGOs

We can increasingly find, within this larger environmental management approach in the water sector, some large global NGOs, such as the World Conservation Union (IUCN). Not surprisingly, these NGOs have close links with the World Bank. As such, they also promote a new process within international development programmes in terms of new kinds of partnership. Indeed, we can now see numerous examples of tri-partnership between the World Bank, the private sector and such global NGOs. WaterAid, for example, works in water and sanitation in partnership with small-scale NGOs and TNCs.[59] Such kind of partnerships are still relatively new, and little has been written about them.

In this section, we have presented the main actors in the international water scene. Indeed, the water sector reflects the whole complexity of international actors, ranging from numerous UN agencies to NGOs and scientific organisations. We have also seen that many of these actions and actors overlap, and that there is a need for co-ordination in the UN, in water and elsewhere. Finally, we have seen that the environmental approach is imposing itself among the international actors around the issue of water resources management. Yet, the fact that the environmental paradigm is imposing itself also means that the World Bank is increasingly taking all relevant actors under its wing. Consequently, the UN actors which are not actively collaborating with the World Bank are being pushed into the direction of humanitarian and security emphasis, while the World Bank is increasingly taking over the area of development, be it in water or elsewhere.

58 Young and Boehmer-Christiansen (1999).
59 See WaterAid (1999).

Conclusion

This chapter has therefore shown how the rising environmental concern – highlighted in Rio and Dublin, and materialised in the so-called Dublin Principles – has brought water issues to the fore, and given rise to corresponding new global actors, such as in particular the WWC, the GWP and the WCD. Environment, indeed, has appeared as the new and main concern since the early 1990s. In the water sector, this means that there is a shift from developmental concerns to seeing water as a resource that has to be properly managed, and in particular managed under economic principles. The pressures and challenges in the global water sector and the corresponding sense of urgency reinforce this call for a managerial and economic approach. In short, there is indeed a need to better allocate, utilise and manage water resources both globally and locally, a claim made particularly strongly by the newly emerging actors such as the WWC and the GWP. All other international actors, in particular those in the UN system, are now increasingly being forced to redefine themselves in light of this new situation and discourse.

In Chapter 1 we saw how the international organisations and institutions are realigning themselves along three topics, namely sustainable development, humanitarian affairs and security, and de-/re-regulation, as well as around three core organisations, namely the World Bank, the UN and the WTO. Environment in general, and water in particular, are clearly falling into the sustainable development topic, and thus into the larger area of influence of the World Bank. As a matter of fact, as shown especially on pp. 49–50, the international organisations dealing with water are gradually being forced to choose between a humanitarian health-centred approach on the one hand, and an environment and development-centred approach on the other. Most of them, including the scientific organisations, are choosing the latter and thus increasingly coming to work with the World Bank, which is imposing itself as the major actor in the water sector. The only other international organisation, which could have played this role (i.e. UNESCO) is rapidly falling behind.

In Chapter 3 we will address the result of these two trends combined: indeed, under the auspices of the World Bank, sustainable development in the water sector basically becomes a question of water resources management. Moreover, the World Bank seems to stress in particular Dublin Principle 4, which recognises water as an economic good. Its economists have transformed the environmental concern into a question of resources management and corresponding investments. Consequently, they foster new partnerships between the World Bank, international NGOs and the private sector. The fact that the UN organisations, given their fragmentation, have not been able to come up with a comprehensive approach to water, as well as the fact that its environmental organisations (i.e. UNEP) are hardly active in the water sector, has further confirmed the World Bank's leadership in global water issues. In short, the World Bank is today the actor driving the change in the global water sector. It is also the World Bank which decides with which partners the global water crisis will be solved in the future – i.e. mainly actors from the private sector, plus, occasionally, some UN agencies and NGOs. This is why we now turn to discussing and presenting the World Bank and its role in global water.

Chapter 3

The World Bank's role and policy in water management

> My observation is that the direction the World Bank wants to take is pragmatic in that it is politically impossible to introduce the unfamiliar economic and environmental principles, which Bank staff believe to be appropriate. Currently they push privatisation more strongly than integrated river basin management but nudge whichever they can according to circumstances. Their main wish is to introduce economic principles to the political processes of allocating and managing water.
>
> (Tony Allan, Professor at the School of Oriental and African Studies, 18 February 2000)

Having presented the main relevant actors involved in global water management, we will now examine, in particular, the role of the World Bank. The World Bank is without doubt one of the most, if not the most important actor in the global water sector, be it in terms of financial aid or in terms of general policy-making in the developing countries. Indeed, 'by the end of the 1991, more than $19 billion had been lent for irrigation and drainage, $12 billion for water supply and sewerage, and about $9 billion for hydropower projects, the total representing more than 15 percent of total Bank lending' (World Bank, 1993: 65–6). The World Bank currently supervises eighty-six water and sanitation projects with loans totalling $5.3 billion.[1]

Within the World Bank structure, several organisational units are involved in water. The three most active ones are (1) the Water Supply and Sanitation Division – which is divided between a Rural Water and Sanitation Programme and an Urban Water and Sanitation Programme, (2) the Global Water Unit[2] and (3) the Private Sector Development Division. It is important to note that the Public Sector Department is not involved in any programmes of institutional reform in the water sector. Some dimensions of water still fall under

1 <http://www.worldbank.org/html/fpd/water/projectlending.html>.

2 The difference between the Water and Sanitation Department and the Global Water Unit is that the first focuses mostly on utility management in the water sector, while the second is responsible for the general management of water resources.

other units and, not surprisingly, some specific water issues are divided up into different organisations and organisational units: hydropower, for example, is divided between the World Bank, which finances projects, and the World Commission on Dams (WCD), which establishes the policy guidelines. Irrigation is under the responsibility of the Agriculture and Natural Resources Department, and the environmental dimensions of water resources management (which includes freshwater, as well as coastal and marine resource management) are divided into several Departments. The Latin America and the Caribbean country Departments, and to a lesser extent, the Middle East and North Africa region, also play a role, owing to the important changes in these regions in recent years in water resources management strategies, and the steep increase in privatisation of the water distribution industry. Moreover, the World Bank is also starting to get involved on issues of international inland waters and conflict resolution.[3] This is not to mention the different partnerships of the World Bank with other organisations, such as NGOs or water TNCs, or, for example, the UNDP/World Bank Sanitation Programme discussed in Chapter 2.

All these organisational units combine to constitute the World Bank's multi-faceted approach to water. Perhaps in response to the need of having some organisational coherence, the concept of 'water resources management' was introduced in a 1993 World Bank Policy Paper (World Bank, 1993). But, when introducing this concept, the World Bank was certainly also aware of the trends described in Chapters 1 and 2, namely the environmental trends calling for an integrated approach to water on the one hand, and growing international organisational realignment on the other. This chapter will therefore examine the process leading up to the formulation of the concept of 'water resources management'. In doing so, we will distinguish three important factors which have contributed to this concept – environment, internal World Bank reform and globalisation. The impetus for a change in global water management, as we have seen in Chapter 2, came essentially from an environmental concern. It is in this respect that the World Bank has integrated the environment into its overall approach, as demonstrated by the 1992 *World Development Report*, addressing the relationship between environment and development. Moreover, recall that in January 1993 a new Vice Presidency for Environmentally Sustainable Development (ESD) was created, precisely in order to incorporate the concept of sustainable development into all aspects of the World Bank's work. Water was the major focus of this reform. The second factor was the necessity of World Bank internal reform in dealing with the management of water resources, due in particular to the failures of the different World Bank's projects of the early 1990s. As we will see in

3 On this subject, see: Kirmani and Rangeley (1994). See also: Salman (ed.) (1999); Salman Salman and Boisson de Chazournes (eds) (1999).

this chapter, this internal factor further accelerated the need for change. The third factor pertains to the developments in Chapter 1: indeed, economic globalisation has, as argued, affected the status of the nation-state, a process accelerated further by neo-liberal economic theory. This point is of particular interest, since the World Bank's new vision on infrastructure marks a complete shift from its previous policies, especially when it comes to the role of the State. Indeed, as we shall see, the World Bank has clearly started to consider that governments should not be involved in infrastructure at all.

The World Bank tries to integrate these three factors into the concept of 'water resources management'. In doing so, it ends up with a serious simplification: privatisation becomes the answer to environmental concerns. We will develop this simplification in five steps: in a first section, we will present the World Bank's role in water up to the 1990s. We then will show how, starting in 1994, its policy towards infrastructure in general and water in particular changed, above all in respect of the role the state is supposed to play. In the third section we will introduce the concept of 'water resources management', which is the key concept integrating both the idea that water is an economic good and the idea that water and other infrastructures need to be privately owned and managed. In the fourth section, we will see how this purely economic approach was somewhat 'softened' by the World Bank during the 1990s, when adopting what they called the 'French model'. In a final section, we will present and critically discuss, the translation of this French model into various practices of water resources management in developing countries in North Africa (Morocco), the Middle East (Yemen) and Asia (Bangladesh).

The World Bank's role in the water sector up to 1990

The World Bank's role in helping member countries – more precisely, developing and Central and Eastern European (CEE) countries – in the water sector is not new. In fact it is rooted in the World Bank's very mission to finance infrastructure projects, which developing countries could not finance by themselves. Consequently, the World Bank has been an important actor for many of the developing countries since the 1950s, investing in different projects pertaining to irrigation, water supply, sanitation, flood control, and hydropower. Figure 3.1 shows that the projects currently supervised by the World Bank in the water sector represent 11 per cent of the Bank's total infrastructure investments.

As compared to other sectors, the water sector therefore represents only a small part of World Bank lending. Tables 3.1 and 3.2 show that water financing boomed in the 1970s (i.e. under Robert McNamara's leadership and fight against poverty), only to decrease thereafter.

However, all these projects, aimed at providing essential services, did in fact not take into account the environmental costs, i.e. the decrease in the

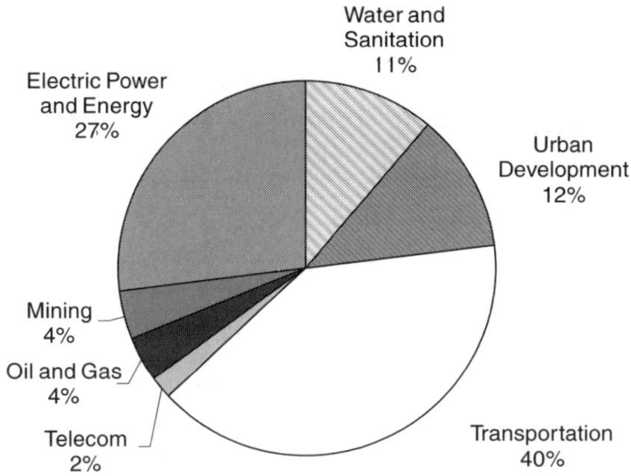

Figure 3.1 World Bank's total infrastructure investments, 2000.
Source: <http://www.worldbank.org.html/fpd/water/projectlending.html>.

Table 3.1 Distribution of World Bank commitments, Financial Year 1968–81 (water supply and sewerage, plus total) (million current $US)

Sector	*Through*	*FY69–73*	*FY74–8*	*FY79*	*FY80*	*FY81*
Water supply and sanitation	130	589	1,329	1,019	631	535
Total	13,250	12,849	32,320	10,011	11,482	12,291

Source: World Bank (1981): 12.

Table 3.2 Distribution of World Bank commitments, Financial Year 1990–9 (water supply and sewerage, plus total) (million current $US)

Sector	*FY90–4*	*FY95*	*FY96*	*FY97*	*FY98*	*FY99*
Water supply and sanitation	1,031.0	981.5	609.8	682.8	476	645
Total	21,925.0	22,521.7	21,516.6	19,146.7	28,594	28,994

Source: World Bank (1999): 10.

availability of water resources, growing water scarcity and growing water stress. From 1992 onwards, as we saw in Chapter 2, this problem became of growing concern for the international community, and the World Bank decided to follow suit (World Bank, 1993). In addition to this awareness of growing water scarcity and stress, the World Bank also came to acknowledge particular problems in the management of its past projects. As highlighted in the reports of the Operations Evaluations Department (OED), the investments supported by the Bank have often encountered implementation, operational and social problems, which were not really addressed by the projects themselves.[4] Underlying these problems is a vicious cycle of poor-quality and unreliable services, resulting in consumers' unwillingness to pay which, in turn, generates inadequate operating funds and a further deterioration in the quality of services.

Indeed, according to the OED 1988 review of approximately 234 water projects for the period between 1974 and 1988, it was found that 88 per cent of the investment in water supply and sanitation and 80 per cent of those in irrigation were satisfactory (World Bank, 1993: 67). This compares with 81 per cent for all Bank projects reviewed. In general, the projects evaluated have pertained to the provision of drinking water and hydroelectric power for some of the world's poorest cities, as well as irrigation projects in order to improve food security and farm family incomes (World Bank, 1993: 67). However, project ratings have dropped sharply for the projects completed in recent years. The percentage of irrigation projects rated satisfactory dropped to 44 per cent during 1989–90, before rising to 71 per cent in 1991. A decline also occurred in water supply and sanitation projects during 1990–1, where only 56 per cent of the projects were judged to be satisfactory (World Bank, 1993: 67). As a result of these internal evaluations, but also as a result of the overall climate which was increasingly concerned with environmental issues (McCully, 1997a), as well as because of popular protest (Cavanagh *et al.*, 1996), the World Bank decided in 1992 to change its policy in the water sector by bringing both environmental and social considerations to bear on the feasibility of its projects (World Bank, 1993). This new way of thinking brought about the concept of 'sustainable water resources management' (World Bank, 1993). This new concept offers a more global vision of water resources and regroups the different World Bank projects (irrigation, water supply, sanitation, flood control and hydropower) into one single concept when dealing with the water sector.

But the challenges due to increased water scarcity and stress, as underlined in Chapter 2, the past failures in the different water projects and the need to bring some internal organisational coherence to the World Bank's water activities, were in fact just one of the two reasons why the concept of 'water resources management' emerged. The other can be found in an equally deep cultural transformation of the World Bank's thinking, namely the shift from Keynesian economics to neo-liberal economics after the second half of the 1980s. As a result – and on top of the fact that the World Bank both enlarged and changed its initial mission – the role of the State in infrastructure services

4 World Bank *Water Supply and Sanitation Projects the Bank's Experience: 1967–1989*, Operations Evaluation Department, Sector Study 10789, 06/19/92, Intranet OED, 1999, § 36.

was entirely being re-evaluated.[5] In this context, the 1994 *World Development Report* created a new framework for the development of infrastructures, of which the water sector was and still is of course fully part.

The World Bank and infrastructure

The World Bank's main role and initial mission is in infrastructure development, thereby financing the vital infrastructures in developing countries. From the point of view of economic theory, such infrastructure financing and development was mainly grounded in Keynesian and classical economics, whereby the nation-state should play not only in macro-economic stability, but whenever market failures occur. And such market failures, according to classical economics, occur quite often in the infrastructure sectors (natural monopoly, externalities, public goods). However, this changed considerably after Robert McNamara quit the World Bank presidency in 1981. Indeed, along with Ronald Reagan as new US president and corresponding new leadership in the World Bank, came a new economic doctrine, namely neo-classical economics. Unlike classical economics, neo-classical economics seeks to reduce the public sector in general and the role of the state in society in particular. Consequently, the World Bank's policy on infrastructures changed. In the Foreword of the *1994 World Development Report*, for example, Lewis T. Preston, the then President of the World Bank, clearly indicated what the Bank's new priorities in infrastructure development would be from now on, when he wrote:

> Several trends are helping to improve the performance of infrastructure. First, innovation in technology makes more diversity possible in the supply of services. Second, an evaluation of the role of government is leading to a shift from direct government provision of services to increasing private sector provision – and recent experience in many countries with public–private partnerships is highlighting new ways to increase efficiency and expand services. Third, increased concern about social and environmental sustainability has heightened public interest in infrastructure design and performance.
>
> (World Bank, 1994d: Foreword)

This quote highlights three factors influencing infrastructure development (in developing countries and elsewhere), namely technology, the role of the

5 See the Foreword from World Bank (1994d): 'Most of this report shows how the functions of the State need to be changed in infrastructure services according to the World Bank'. This idea was then further developed in the 1997 *World Development Report*, entitled *The State in a Changing World* (World Bank, 1997c).

state and social and environmental concerns. We have addressed the last in Chapter 2, and seen how these have been a significant factor in pushing towards a more economic approach in the provision of public services in the water sector internationally. We will not address the technological dimension of public service provision in this book. However, in this section we would like to examine in more detail two aspects of this changing role of the state in infrastructure. First, we will analyse how the World Bank defines the new role of the state in infrastructure development and service provision. We will then present the alternative mechanisms to the state for such infrastructure development and service provision, mechanisms, which can be subsumed under the heading of 'public–private partnerships'.

Redefining the role of the State in infrastructure

In its 1994 *World Development Report*, the World Bank defined a new approach to infrastructure development and service provision in general. In particular, and most certainly under the influence of neo-classical economics, the World Bank reconsidered the role of the state. Unlike previously where the state had been a crucial factor in infrastructure development, it had become an impediment to 'progress':

> In developing countries, governments own, operate and finance nearly all infrastructures, primarily because its production characteristics and the public interest involved were thought to require monopoly – and hence government – provision. The record of success and failure in infrastructure is largely a story of government's performance. Infrastructure's past growth has in some respects been spectacular . . . In other important respects, however, the performance has been disappointing. Infrastructure investments have often been misallocated – too much to new investment, not enough to maintenance; too much to low priority projects, not enough to essential services. The delivery of services has been hampered by technical inefficiency and outright waste. And too few investments and delivery decisions have been attentive to meeting the varied demands of different user group, or to the consequences for the environment.
>
> (World Bank, 1994d: 4)

Note that the World Bank uses the term 'government' rather than 'state'. This changes only after its 1997 report entitled *The State in a Changing World*, from whence the words 'government' and 'state' are used interchangeably. But in any case, the Bank identifies three main reasons for such poor performance, all three stemming from the fact that government (or the State) is involved in the provision of services:

> First, the delivery of infrastructure services usually takes place in a market structure with one dominating characteristic: the absence of competition. Second, those charged with the responsibility for delivering infrastructure services are rarely given the managerial and financial autonomy they need to do their work properly . . . They are compelled to deliver services below cost – often by not being allowed to adjust prices for inflation . . . The other side of the coin is that public providers are rarely held accountable for their actions. Third, the users of infrastructure – both actual and potential – are not well positioned to make their demands felt . . . Excess consumer demand based on below-cost prices is not a reliable indicator that services should be expanded, although often it is taken as such.
>
> (World Bank, 1994d: 7)

In the water sector, for example, the past failures had two different reasons, according to the World Bank's private sector specialists,[6] namely wrong (political) priorities and politics. The first error was governments' failure to maintain infrastructure networks. Indeed, most governments focused on expanding services, rightly assuming that water should be provided for all. The World Bank responded that maintenance, rather than the development of infrastructure, should become a priority. Consequently, the World Bank went into the training of the water utilities' staff, so as to bring about better ways of dealing with the management of water resources. However, this appeared to be a second failure, and the World Bank concluded that any public service was always under certain political constraints, and therefore not able to make as rational decisions as would a World Bank economist.

Consequently – i.e. because of governments' inability to properly manage infrastructures – the World Bank changed its theory and its practice. Moreover, it argued, this change was made possible thanks to three converging forces, namely technology, neo-classical economic theory and environmental pressures: 'First, important innovations have occurred in technology and in the regulatory management of markets. Second, a consensus is emerging on a larger role for the private sector in infrastructure provision, based in part on recent experience with new initiatives. Third, greater concern now exists for environmental sustainability and for poverty reduction' (World Bank, 1994d: 7).

In essence, the World Bank increasingly refuted the idea that infrastructure services are public goods, which thus would warrant that the state should play a more active role in their provision. Indeed, though it admits that this might have been the case earlier, the Bank now tends to consider that infrastructure

6 Personal interview conducted at the World Bank on 20 March 2000.

services can be treated (though not perfectly) as private goods. In the case of water, for example, the logic of the Bank goes as follows:

> It has been common in many countries not to charge users for the volume of some utility services consumed because the marginal supply cost was considered negligible, congestion was absent, or technological constraints (such as the absence of water meters) prevented volume pricing. However, recent developments, such as the increased scarcity (and supply cost) of water, growing congestion as network capacity becomes fully utilized, and technical innovations in metering consumption, have made it possible and desirable to price these services like other private goods.
> (World Bank, 1994d: 23)

Yet, this is not exactly true, as the World Bank itself must admit that in some cases full cost pricing may not be acceptable, meaning that water, for example, cannot be considered a totally private good. As the Bank says just one page later: 'Certain characteristics of infrastructure also create challenges in financing. Where a minimum level of consumption of a particular service (such as water, heating or power) can be identified as a "lifeline" for some users, society may judge that they [the users] should not be excluded from infrastructure they cannot afford to pay' (World Bank, 1994d: 24). This indeed will create, according to the World Bank, a problem in financing water and other infrastructure services, as no private operator is willing to deliver its services below cost. Thus, water is like any private good, yet at the same time special. Incidentally, we can find the same ambiguity among the companies delivering such services, as they also want to be perceived as something special, thus calling themselves 'Public Service Enterprises' (see Chapter 4). Given this ambiguity, the World Bank proposes the three following compromise measure when it comes to infrastructures services provision in general and water provision in particular, i.e. 'the wider application of commercial principles to service providers, the broader use of competition, and the increased involvement of users where commercial and competitive behaviour is constrained' (World Bank, 1994d: 8). Note that 'user participation' is warranted only when market mechanisms fail.

Two main arguments are given by the Bank as to why such market mechanism (i.e. both commercial principles when running the infrastructure companies and competition) will prove successful. First, it is said, this will reduce the financial burden on the state. Indeed, the World Bank estimated that, 'for only three sectors (power, water and railways), the total savings are nearly $123 billion annually – nearly 10 percent of total government revenues in developing countries, 60 percent of annual infrastructure investment, and approximately five times annual development finance for infrastructure' (World Bank, 1994d: 11). Such savings are basically due to efficiency gains. Secondly, the World Bank argues, reforming infrastructure development and service provision along market mechanism lines will improve quality of service, as well as national economic competitiveness:

> Improvements in productivity and pricing would permit more effective delivery of service in response to demand. They would also enhance the growth and competitiveness of the economy. And they would allow vastly greater mobilization of resources for needed new investments by generating higher revenues and by creating a policy environment conducive to the inflow of new investment resources.
>
> (World Bank, 1994d: 11)

In short, Private Sector Participation (PSP) in infrastructure development and services provision will not only bring new technologies and a managerial approach, thus making networks more efficient, but it will also provide water to those peoples who presently have no access. Commercial principles, competition and private sector participation will thus, says the Bank, solve two problems in one: the state will have better maintained infrastructures and a more extended network reaching more people. Let us now see how this works in practice.

Alternative mechanisms for providing water

There are four ways to improve upon the current delivery of water services, according to the World Bank, namely (1) public ownership yet operated by an enterprise or a department, (2) public ownership with operation contracted out to the private sector, (3) private ownership and operation, often accompanied by (monopoly and maybe other) regulation and (4) community and user provision (World Bank, 1994d: 8). With the exception of the last option, these are actually standard in the literature. Let us present here the first three options of delivering water services, and show that for the World Bank private ownership and operation is the optimal solution.

The first option is in fact the application of some basic management principles to the way public services have traditionally operated. More precisely, the World Bank suggests the application of three such principles, namely corporatisation, contracts, and pricing:

- Corporatisation, which establishes the quasi-independence of public entities and insulates infrastructure enterprises from non-commercial pressures and constraints;
- explicit contracts between governments and (public or private) managers or private entities involved in infrastructure services, which increase autonomy and accountability by specifying performance objectives that embody government-defined goals; and
- a pricing strategy designed to ensure cost recovery, which creates a

> desirable form of financial independence for public utilities and even at times for public works.
>
> (World Bank, 1994d: 37)

This first option, according to the World Bank, is not optimal, since it does not offer sufficient opportunities for profit-making. The argument, here, however, is quite tautological: 'Although effective in the long run for some sectors and some services, this solution raises at least two problems. First, and more obviously, providers in many cases are in the public sector precisely because of the limits on profit maximization, either because the services are public goods (as with roads) or because governments have objectives other than profit' (World Bank, 1994d: 41).

The second option does go further than the first, as operators are now private firms, whose aim it is to make a profit. In this case, the World Bank considers three other approaches, namely performance contracts, management contracts, or service contracts:

> - Performance agreements retain all decisions in the public sector. They try to increase the accountability of employees and managers and to improve the focus of operations by clarifying performance expectations and the roles of responsibilities, and rewards of all those involved.
> - Management contracts transfer to private providers the responsibility for managing an operation such as port or a power or water utility. They increase the autonomy of management and reduce the risks of political interference in the day-to-day operations of public entity.
> - Service contracts transfer to private providers the responsibility for delivering a specific service at lower costs or obtaining specific skills or expertise lacking in the public sector – such as design engineering.
>
> (World Bank, 1994d: 42)

Yet again, the World Bank does not believe that any of these measures will bring sufficient changes from the way infrastructures have been developed and managed so far. Indeed, the Bank says: 'Simply establishing commercial principles and maintaining them through political commitment are not sufficient for the success of commercial enterprises, however. The missing element for success is the introduction of competition with appropriate regulation' (World Bank, 1994d: 51).

In other words, nothing but competition will do. Let thus briefly look at what the Bank considers to be 'competition'. In infrastructure, the Bank says, there are three types of competition, i.e. first the competition from substitutes – for example, when electricity users switch to gas – something quite difficult to achieve in the water sector. Secondly, there is competition in the market itself, i.e. when users can choose between products from different companies, say water from different water companies, where all government would do is ensure fair competition – again, something quite difficult to achieve in the water sector, at least at present. Thirdly, there is 'competition for the market', where governments create competitive conditions through leases or concessions, and firms compete not for individual consumers in the market but for the right to supply the entire market. Should this not be possible and (natural) monopolies still persist, the 'transfer to private ownership generally yields efficiency gains' (World Bank, 1994d: 53). The role of the government is then to regulate monopolies by means of some sort of incentive mechanism such as price-cap or rate-of-return regulation. This, according to the Bank, is still better than government operatorship. However, not all economists agree with this affirmation, as monopoly regulation entails all kind of problems (Beesley and Littlechild, 1991; Newbury, 2000: Chapter 2).

In any case, the World Bank considers – and this is the third (i.e. the preferred) option – that private ownership and operation is by far preferable to anything else. Should a monopoly persist, monopoly regulation may be (regrettably in the eyes of the World Bank) necessary. The goal is thus to reduce monopolies as much as possible, while subjecting everything else to market mechanisms, which is a typical consideration of neo-classical economists. Indeed, it is neo-classical economists, who have come up with the idea of 'unbundling', an idea which the World Bank wholeheartedly endorses in infrastructure in general and in water in particular when saying: 'by isolating the natural monopoly segments of an industry, unbundling promotes new entry and competition in segments that are potentially competitive. Failure to unbundle can constrain an entire sector to monopoly provision even when the numerous activities can be undertaken competitively' (World Bank, 1994d: 53).

Figure 3.2 highlights how, in the World Bank's view, the different ways of unbundling can be realised, as well as the consequences this has for industry structure, competition and regulation.

Figure 3.2 also shows that regulation has different functions, depending upon the extent of competition. Let us note here that regulation, for the World Bank, is in essence (economic) regulation of an economic monopoly, thus not regulation of public services provision, for example, which is a regulation that would also purse social and political objectives (Cox, 1999; Eberlein, 1999). The World Bank thus distinguishes between two types of regulation, namely market regulation, which is about fair competition on the one hand and monopoly regulation, which is about prices and quality, on the other. This is not a discussion we can pursue here, but let us simply note that for the World Bank (monopoly) regulation is something that is strictly limited to something that cannot be unbundled, i.e. the ultimate monopoly. Such monopoly regulation is residual, and should disappear over time, as technological progress and market growth make monopolies disappear.

All these considerations, as we will now see, have been incorporated into

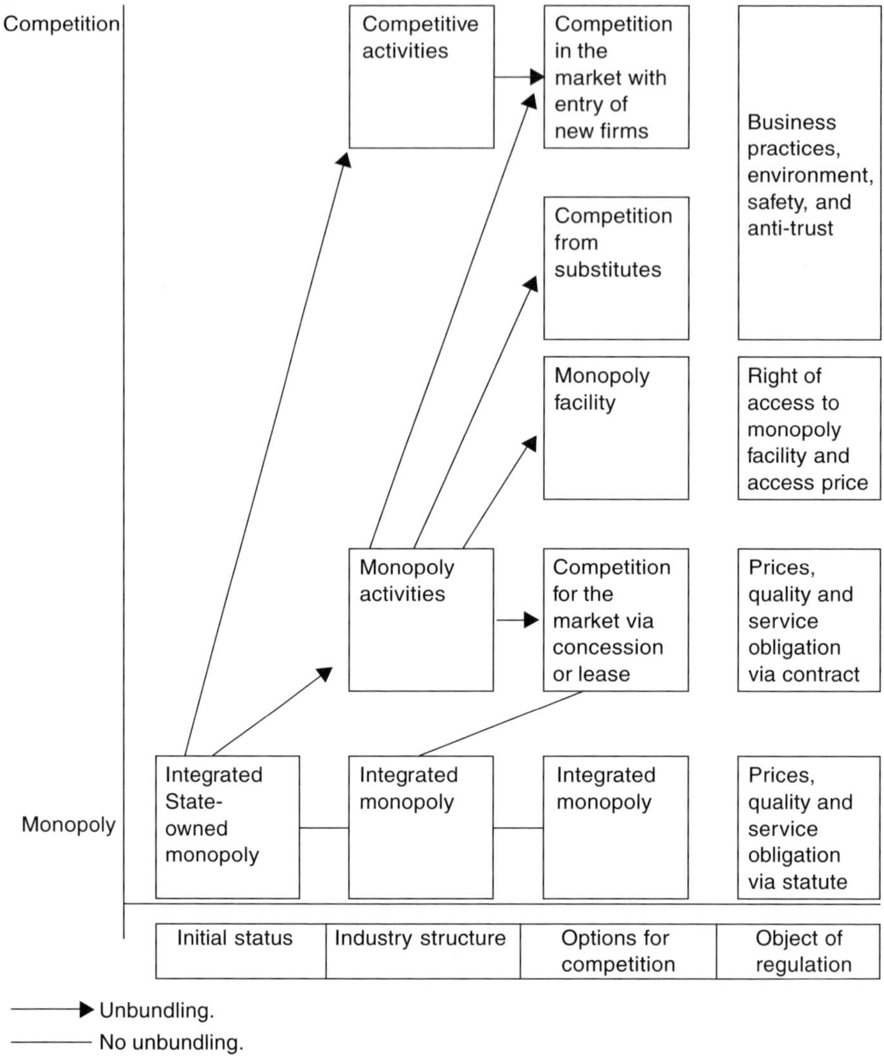

Figure 3.2 Unbundling increasing the options for competition and private sector involvement. *Source:* World Bank (1994d: 56).

the World Bank's concept of 'water resources management'. This concept is thus the combined result of two separate trends: on the one hand the trend to unbundle we have just discussed, and on the other hand the trend we discussed in Chapter 2, namely to consider water as an economic good. Both miraculously seem to reinforce each other and water resources management emerges as the solution where there is no more contradiction between environmental protection and water privatisation.

The concept of sustainable water resources management

The concept of water resources management was introduced in a 1993 World Bank Policy Paper (World Bank, 1993). This paper served as a guideline for the World Bank in developing a new policy grounded in a general theory of water resources management at the local, national and international levels. The following quote indicates the scope of this concept:

> Sustainable growth is predicated on improved water management to overcome constraints posed by the quantity and quality of water available for development. The failure of past practices to adequately deal with the challenges highlights the need for a new approach. The adoption of a program of integrated water resource management holds the key to placing water resources . . . on a sustainable path. Meeting the challenges requires the integration of the qualitative, quantitative, economic and institutional dimensions to water resource management.
>
> (World Bank, 1995a: 42)

This concept of (sustainable) water resources management, in turn, must be seen in the context of the international agreements reached (Rio, Dublin) along with other pressures on the World Bank. Indeed, although the environment was a main concern in formulating this new approach, the Bank also learned from its past failures in project implementations in the water sector, or in the infrastructure sector more generally, as seen above. The Bank's approach is therefore based on the following two guiding principles, namely environmental protection and efficiency. Both seem to go hand in hand, when the Bank says:

> First, water resource management policies and activities should be formulated within the context of a comprehensive analytical framework that takes into account the interdependencies among sectors and protects aquatic ecosystem. Such a framework would guide the establishment of improved co-ordination among institutions, consistent regulations, coherent policies, and targeted government actions. Second, efficiency in water management must be improved through the greater use of pricing and through greater reliance on decentralization, user participation, privatization and financial autonomy to enhance accountability and improve performance incentives.
>
> (World Bank, 1993: 40)

Water resources management as a concept replaces 'water resources development', which is judged to be unsustainable. The following quote from a corresponding World Bank project in Bangladesh perfectly illustrates this underlying idea:

> In Bangladesh, water resource development during the lean season has been geared toward the expansion of irrigation to promote food-grain production. With conflicts arising from competing uses of even more limited supplies of water, the time has come to shift the focus from water resource development to water resources management. Population growth, urbanization, and industrialization will not allow continuation of current practices. If they do continue, they will hurt the economy and cause permanent damage to the environment.
>
> (World Bank Bangladesh, 1998: Chapter Three)

In short, in order to promote an environmentally and economically sustainable development, water resources have to be 'managed'. Such management is possible only if water becomes an economic good, an idea already present in the Dublin principles. In other words, the underlying assumption of water resources management is that water is an economic good. And this really constitutes the main change in the policy adopted by the main World Bank's water specialists. Indeed, if water was an economic good, people and politicians would behave more responsibly, and unnecessary water wasting worldwide could be prevented. The World Bank thus seeks to develop a general policy and approach, which will lead to a more efficient use of water worldwide, but above all in the developing countries. This approach is called 'sustainable water resources management'. Besides being grounded in the Dublin Principles, especially Principle 4, it is also significantly inspired by the ongoing reforms in infrastructure development as outlined above. In this regard, it is the World Bank's aim to reform a country's institutional and legal environment in order to empower water users to make their own decisions regarding the use of the resource, while at the same time providing a structure that reveals the real scarcity value of water. 'Water as an economic good', 'privatisation', 'decentralisation', and user participation thus emerge, according to the World Bank, as the solution to the world's water problems. In this section we will thus present the theory and the practice of water as an economic good, as well as the World Bank's ideas on privatisation, decentralisation, and user participation.

The theory of considering water as an economic good

As underlined in a paper prepared for a meeting of the Technical Advisory Committee of the Global Water Partnership in Namibia, there is substantial confusion about what it exactly means to treat water as an economic good (Rogers *et al.*, 1998: 4). According to Dublin Principle 4, 'water has an economic value in all its competing uses and should be recognised as an economic good'. However, neither the Dublin Principles, nor Agenda 21 say what this means.

On a more theoretical level, the definition of water as an economic good appears to be quite simple. Says John Briscoe, one of the World Bank's senior water advisors: 'like any other good, water has a value to users, who are willing to pay for it. Like any other good, consumers will use water so long as the benefits from use of an additional cubic meter exceed the costs so incurred' (Briscoe, 1996: § 1). Other authors put the same idea into a more technical language:

> general principles . . . in assessing the economic value of water and the costs associated with its provision [are]: First, an understanding of the costs involved with the provision of water, both direct and indirect, is key. Second, from the use of water, one can derive a value, which can be affected by the reliability of supply, and by the quality of water . . . Regardless of the method of estimation, the ideal for the sustainable use of water requires that the values and the costs should balance each other; full cost must equal the sustainable value in use.
>
> (Rogers *et al.*, 1998: 5)

More precisely, several factors must be considered when calculating the cost of water. John Briscoe, in his paper, limits the costs to use cost and opportunity cost. The 'use cost' pertains to the construction and operation of the infrastructure necessary for storing, treating and distributing water. The 'opportunity cost' defines the costs incurred when one user uses water, and therefore affects the use of the resource by another user (Briscoe, 1996: § 3). However, this definition fails to consider other factors, which are equally important. As underlined by other authors, both economic and environmental externalities need also to be considered, such as for example externalities 'associated with the impact of an upstream diversion of water or the release of pollution on downstream users. There are also externalities due to over-extraction from, or continuation of, common pool resources such as lakes and underground water' (Rogers *et al.*, 1998: 8).

The full cost of water is therefore the full economic cost plus the environmental externalities. 'Environmental externalities' are those costs associated with public health and ecosystem maintenance. In a situation of an ideal sustainable use of water, the full cost must equal the sustainable value in use. Figure 3.3 summarises this idea.

The practice of considering water as an economic good

If we refer to the above definition, considering water as an economic good seems to be a comprehensive approach, which takes even environmental and social factors into account. However, this approach is not without problems, especially in developing countries. We will discuss here the practical problems of costing water in urban settings, in irrigation as well as for the poor. In doing so, we will basically refer to a keynote address given by John Briscoe to the

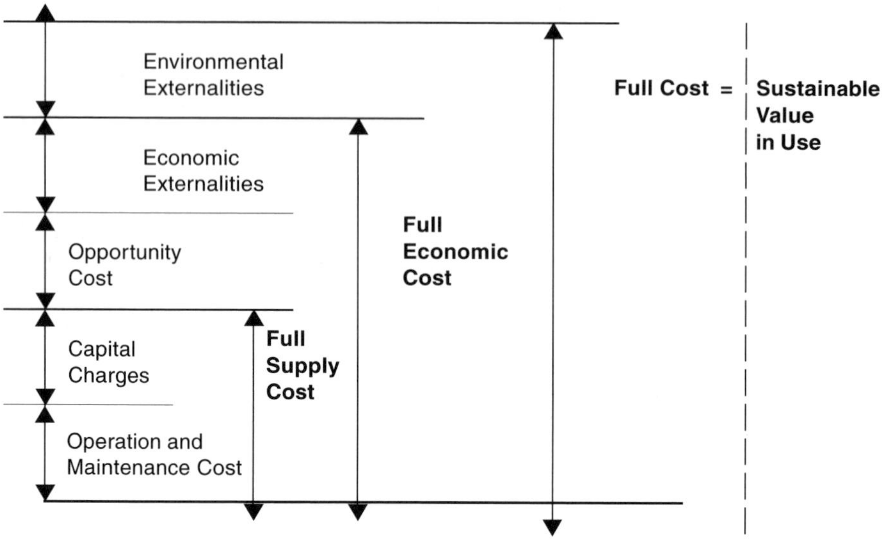

Figure 3.3 General principles for calculating the cost of water.
Source: Rogers *et al.* (1998: Figure 1).

International Committee on Irrigation and Drainage Conference on Water as an Economic Good (Briscoe, 1997).

In urban water supply for example, the supply costs are relatively high, while the opportunity costs are quite low. If we consider water as an economic good, users should be charged full marginal costs. The World Bank's solution to this problem consists of gradually increasing user chargers starting from an average financial cost, which includes the 'historical costs'. The historical cost is the original investment for the service plus the future replacement costs of the facilities. As we will see, the original investment is often quite difficult to calculate. Furthermore, other problems must also be considered. For example, 'many cities in developing countries are growing rapidly . . . incomes are also increasing and industrial demand is growing very fast and new sources have constantly to be found. A consequence is that the costs of urban supplies from new sources are growing rapidly' (Briscoe, 1996: § 4.2). This dynamic is real, as shown in the case of the costs for raw water in urban areas in World Bank-financed projects, where it was found that the unit cost of water can more than double, or triple in some cases (World Bank, 1992a: 101). Moreover, urban water supplies in most countries have been financed out of general revenues. In many cases, these costs are fully subsidised, with the utility responsible only for operation and maintenance costs. The elimination of such subsidies should of course occur gradually. This brings us to the question, unanswered as yet by the World Bank, of how to reconcile private sector participation, full cost pricing, and state subsidies? As we shall see later, state subsidies have in fact in most cases been replaced by World Bank loans.

There are also many problems in applying this approach to *irrigation* in developing countries. In irrigation, the supply cost is usually low, while the opportunity cost is very high. However, in reality opportunity cost are usually not charged, or charged at a very low price. But charging full opportunity costs would almost be impossible, as, for example farmers in Chile, Australia, or even California would be asked to pay more than ten times the current price of water (Briscoe, 1997: § 1.2). Moreover, many people are concerned by the fact that treating water as an economic resource will mean the end of irrigated agriculture. This, in turn, would create a problem for food production. Indeed, a report by the International Food Policy Research Institute entitled *Water Resources in the Twenty-First Century: Challenges and Implications for Action*, shows that agriculture is by far the biggest user of water, accounting for more than 70 per cent of total water consumption in poor developing countries.[7]

Another obvious problem is the impact such an approach would have on the *poor*, since water, so far, has been heavily subsidised. There again, senior officers of the World Bank like John Briscoe argue that the recent experiences of the urban water sector do in fact show the opposite: 'The poor are much better off when water is managed as an economic good' (Briscoe, 1997, § 3). Indeed, according to the World Bank, increasing prices to enable cost recovery in the delivering of services may actually help the poor. The Bank's argument here is that the poor often pay much higher prices per unit for privately provided water because they are not connected to public service networks:

> This effect had been demonstrated most convincingly for water, where the concerns for the poor are properly strong. In the Brazilian city of Grande State of Vitoria, Esperito Santo, the willingness to pay for new water connections in 1993 was four times the cost of providing the service, while the willingness to pay for sewerage collection and treatment was 2.3 times the cost. Without treatment before disposal, the willingness to pay falls to only 1.4 times the cost because untreated sewage creates health problems and reduces the recreational value (mostly the fishing yield) of the waters into which it is discharged.
>
> (World Bank, 1994d: 49)

Moreover, the Bank considers that easier access to water can free up time, which can be used to pursue income-earning activities.

7 'Report finds world water supplies dwindling while demand rises; world food production, health, and environment at risk' (9 March 1997) <http://www.cgiar.org/ifpri/pressrel/2030997>.

The above shows that, in practice, defining water as an economic good is not without its problems. Practical use of this approach might thus well be quite limited. The World Bank acknowledges this, but argues that water as an economic good must be seen and used in association with the three other concepts, which are decentralisation, privatisation and user participation. Only together can these make up for the World Bank's new idea of (sustainable) water resources management. Let us thus examine each of these three in more detail.

Privatisation

The main problem in the water sector, as in most infrastructure sectors, is its lack of financial autonomy. This problem is, indeed, particularly acute in the water sector, where in developing countries on average 70 per cent of the running costs are financed by government subsidies. (World Bank, 1994d: 47). At the same time, and despite the progress made in developing countries over the past fifteen years, the share of households with access to clean drinking water has increased by only 50 per cent (World Bank, 1994d: 1). As a consequence, 1 billion people still lack access to clean water and nearly 2 billion lack adequate sanitation (World Bank, 1994d: 1). The correlation between these two facts – i.e. subsidies and development (or the lack thereof) – is at the core of the World Bank's water policy reform. Indeed, the Bank wants to promote drinking water and sanitation, while at the same time reduce the role of the State along the logic outlined on pp. 73–4 of this chapter, i.e. by means of unbundling. If we recall, the Bank's idea is to unbundle whatever can be handed over to the private sector, which means, in essence, privatisation.

The lack of progress in the water sector is thus not blamed on the sector's specificity, but on government failure. A striking example can be found in the 1994 *World Development Report*, which estimates that for public water utilities in Latin America, annual financial losses represent 15 per cent or more of the investments needed to supply the entire population with adequate services by the turn of the century (World Bank, 1994d: 47). As a consequence,

> the best way of reducing the gap between costs and revenues is to cut costs and achieve productive efficiency – perhaps the most important lesson of the Bank's experience in infrastructure. Costs due to the poor debt management are excessive in about one-third of World Bank-supported infrastructure projects. Maintenance problems that cause water or power losses are even more common and costly. In Costa Rica, the national water company estimates an annual loss of income from such losses equivalent to 24 percent of investment planned for the next five years. In Mexico city, at the end of the 1980s, neglect of maintenance and the lags

> between tariff increases and costs increases in the water sector required a federal subsidy amounting to about 0.6 percent of GDP a year.
>
> (World Bank, 1994: 47–8)

It is against this background that privatisation appears, from the point of view of the World Bank, to be the best solution.

But privatisation is not good only for efficiency reasons, it is also, according to World Bank specialists, beneficial for solving environmental problems. Todd Hanson and Letitia Oliveira of the International Finance Corporation's (IFC) Technical and Environment Department go as far as to say that 'privatizations can make substantial contributions to sustainable development' (Hanson and Oliveira, 1997: 34). One of the examples they give refers to the privatisation of Manila's water supply system, where 'as a result of privatization, average water tariffs will decrease by more than 50% from present levels, water pressure and availability will meet international standards within three years, universal water coverage will be assured within ten years, and more than 80% of the population will have sewerage and sanitation services within the concession period' (Hanson and Oliveira, 1997: 35).

We thus find here again the same arguments for privatisation as in the second section of this chapter, namely efficiency and environmental sustainability. But, as in the above section, privatisation can take various forms. Rather, we should speak, as does the Bank, of various forms of Private Sector Participation (PSP), or Public–Private Partnership (PPP). In addition, so the argument goes, public funds alone cannot meet the large investment needs of the water sector, both in water and in wastewater services. Box 3.1 presents these different options of private sector participation.

Box 3.1 Different forms of private sector participation

Private firms can be brought into the operation and/or development of water and wastewater services in different ways, which vary by their scope and the extent of responsibilities and risks assumed by the firms and the contracting authority.

- Under a *service contract*, a utility hires a private firm to carry out specific tasks (e.g. leak detection, meter reading, water quality measurements) while retaining full responsibility and risk for service provision. Sub-contracting is especially advisable for activities that have large peaks (for instance, construction), or when the in-house level of activity is insufficient to reach an efficient scale of operation (for instance, water quality laboratory, or vehicle fleet maintenance). Even for other activities (for instance, meter reading) private contractors working under competitive pressure are often more cost-effective than utility departments.
- A *management contract* is a service contract where the utility or its municipal owner hire a management team from a private firm. It can be used to bring in

new management systems, organizations and skills, or as a preliminary step to restructure a dilapidated utility before a concession. Compensation is costs-plus-fee, and the contracting municipality retains most of the operational and commercial risks of the utility, though some risk-sharing may be built into the contract using performance bonuses or contingent fees.

- In a *lease contract*, a private firm takes over the operation and maintenance of the system, collects user charges, and is compensated with an agreed portion of the revenues. The municipality remains responsible for system expansion and replacement of major assets and recovers parts or all of its costs from its own share of user charges. The lease holder may also administer investment funds as agent to the municipality, without taking related financial risks.
- In a *concession* or build–operate–transfer (BOT) scheme, private investors arrange the financing and construction (or rehabilitation) of either a self-contained facility (e.g. a treatment plant) or a complete water/sewerage system, then operate it for the period for the concession. For a treatment plant-type BOT, the investor is compensated by the public utility under a 'take-or-pay' obligation, and the public utility retains the commercial relationship with the end-users and assumes the related risks. In a system-wide concession, the investor bills and retains user charges for the concession period.
- Full *private ownership*, arising from the sale of all utility shares to private investors (the model in England and Wales), is unlikely to materialize in Poland for at least the next few years. Even if municipalities were interested in selling (the proposal would face strong political opposition in most cities), they would not find responsive buyers, given (a) the lack of regulatory experience, and uncertainties on sector institutions, policies, the utility's cost structure and revenue base, and the macroeconomic environment; and (b) the limited size of the capital markets and the lack of private domestic enterprises with relevant experience and sufficient resources.
- A municipality can share ownership with private shareholders in a *joint venture* company. This company itself may either own the assets or (most often) be given a franchise by the local government as in one of the lease or concession arrangements described above. The option is popular in Central and Eastern Europe.

Source: World Bank (1997d: Annex 9).

Decentralisation and user participation

While privatisation, in its different forms, constitutes a core aspect of the World Bank's new policy in water resources management, decentralisation is in fact the underlying framework within which privatisation should be located (Serageldin, 1994). However, as we will show, decentralisation as a concept is in fact very close to privatisation. This can be explained by the fact that, within the World Bank, there seem to coexist two different approaches. One comes from the general reform of the public sector and infrastructures pushing for a greater participation of the private sector as well as for what is called 'market decentralisation'. The other seems to be concerned with sustainable environment, and thus pushes for a more decentralised approach based on river basin management. 'Decentralisation' thus appears as the organisational compromise between these two trends.

In fact, decentralisation is not defined by the World Bank in geographical, but in political terms: 'Decentralization [is] the transfer of authority and responsibility for public functions from the central government to subordinate or quasi-independent government organizations and/or the private sector'.[8] The advantage of such a definition is that it can satisfy two types of critiques of government, namely the critique from the private sector on the one hand and the critique from civil society on the other. Both, to a certain extent, want power to be devolved from the nation-state to more autonomous units. But, as a matter of fact, the Bank distinguishes between four different forms of decentralisation, i.e. political, administrative, fiscal, and market decentralisation, the last being more or less equivalent to privatisation, as shown in Box 3.2.

Box 3.2 Different forms of decentralisation according to the World Bank

Political decentralization

Political decentralization aims at giving citizens or their elected representatives more power in public decision-making. It is often associated with pluralistic politics and representative government, but it can also support democratization by giving citizens, or their representatives, more influence in the formulation and implementation of policies. Advocates of political decentralization assume that decisions made with greater participation will be better informed and more relevant to diverse interests in society than those made only by national political authorities. The concept implies that the selection of representatives from local electoral jurisdictions allows citizens to know better their political representatives and allows elected officials to know better the needs and desires of their constituents. Political decentralization often requires constitutional or statutory reforms, the development of pluralistic political parties, the strengthening of legislatures, creation of local political units, and the encouragement of effective public interest groups.

Administrative decentralization

Administrative decentralization seeks to redistribute authority, responsibility and financial resources for providing public services among different levels of government. It is the transfer of responsibility for the planning, financing and management of certain public functions from the central government and its agencies to field units of government agencies, subordinate units or levels of government, semi-autonomous public authorities or corporations, or area-wide, regional or functional authorities.

8 <http://www.worldbank.org/publicsector/decentralization/Different.htm>.

The three major forms of administrative decentralization – deconcentration, delegation, and devolution – each have different characteristics.

Deconcentration. Deconcentration – which is often considered to be the weakest form of decentralization and is used most frequently in unitary states – redistributes decision-making authority and financial and management responsibilities among different levels of the central government. It can merely shift responsibilities from central government officials in the capital city to those working in regions, provinces or districts, or it can create strong field administration or local administrative capacity under the supervision of central government ministries.

Delegation. Delegation is a more extensive form of decentralization. Through delegation central governments transfer responsibility for decision-making and administration of public functions to semi-autonomous organizations not wholly controlled by the central government, but ultimately accountable to it. Governments delegate responsibilities when they create public enterprises or corporations, housing authorities, transportation authorities, special service districts, semi-autonomous school districts, regional development corporations, or special project implementation units. Usually these organizations have a great deal of discretion in decision-making. They may be exempt from constraints on regular civil service personnel and may be able to charge users directly for services.

Devolution. A third type of administrative decentralization is devolution. When governments devolve functions, they transfer authority for decision-making, finance, and management to quasi-autonomous units of local government with corporate status. Devolution usually transfers responsibilities for services to municipalities that elect their own mayors and councils, raise their own revenues, and have independent authority to make investment decisions. In a devolved system, local governments have clear and legally recognized geographical boundaries over which they exercise authority and within which they perform public functions. It is this type of administrative decentralization that underlies most political decentralization.

Fiscal decentralization

Financial responsibility is a core component of decentralization. If local governments and private organizations are to carry out decentralized functions effectively, they must have an adequate level of revenues – either raised locally or transferred from the central government – as well as the authority to make decisions about expenditures. Fiscal decentralization can take many forms, including (a) self-financing or cost recovery through user charges, (b) co-financing or co-production arrangements through which the users participate in providing services and infrastructure through monetary or labor contributions; (c) expansion of local revenues through property or sales taxes, or indirect charges; (d) intergovernmental transfers that shift general revenues from taxes collected by the central government to local governments for general or specific uses; and (e) authorization of municipal borrowing and the mobilization of either national or local government resources through loan guarantees. In many developing countries local governments or administrative units possess the legal authority to impose taxes, but the tax base is so weak and the dependence on central government subsidies so ingrained that no attempt is made to exercise that authority.

Economic or market decentralization

The most complete forms of decentralization from a government's perspective are privatization and deregulation because they shift responsibility for functions from the public to the private sector. Privatization and deregulation are usually, but not always, accompanied by economic liberalization and market development policies. They allow functions that had been primarily or exclusively the responsibility of government to be carried out by businesses, community groups, cooperatives, private voluntary associations, and other non-government organizations.

Privatization. Privatization can range in scope from leaving the provision of goods and services entirely to the free operation of the market to 'public–private partnerships' in which government and the private sector cooperate to provide services or infrastructure. Privatization can include: (1) allowing private enterprises to perform functions that had previously been monopolized by government; (2) contracting out the provision or management of public services or facilities to commercial enterprises (indeed, there is a wide range of possible ways in which function can be organized and many examples of within public sector and public–private institutional forms, particularly in infrastructure); (3) financing public sector programs through the capital market (with adequate regulation or measures to prevent situations where the central government bears the risk for this borrowing) and allowing private organizations to participate; and (4) transferring responsibility for providing services from the public to the private sector through the divestiture of state-owned enterprises.

Deregulation. Deregulation reduces the legal constraints on private participation in service provision or allows competition among private suppliers for services that in the past had been provided by the government or by regulated monopolies. In recent years privatization and deregulation have become more attractive alternatives to governments in developing countries. Local governments are also privatizing by contracting out service provision or administration.

Source: <http://www.worldbank.org/publicsector/decentralization/Different.htm>.

The fact of putting all these four aspects under the general heading of 'decentralisation' can indeed help to reconcile the two different trends within the Bank, namely the more market-oriented trend which seeks to promote private sector participation and ultimately privatisation with the more environmentally and civil society-oriented trend. Nevertheless, one can see that there clearly is a hierarchy, which puts economic or market decentralisation above the three others, and ends up redefining and reducing civil society concerns to 'user participation' (see pp. 86–7). The term 'decentralisation' also helps hide the fact that what we are talking about here is ultimately reform and not decentralisation. Thus one should rather read 'Political reform', 'administrative reform', 'fiscal reform' and 'economic reform'.

In short, the concept of decentralisation, in this case in the water sector, has the particular advantage of combining four types of reform, namely political, administrative, fiscal and economic. More precisely, it can reconcile 'privatisation' (economic and fiscal reform) with 'user participation' (political and administrative reform). As the Bank says:

> This new view entrusts consumers and neighbourhoods with a large say regarding the services they want and how to pay for them. It therefore gives the actual users much more choice, as well as responsibility, than did the old top-down approach. It means that the poor – who are often left out of theoretically universal arrangements that mainly benefit and subsidize non-poor users – can gain access to the services they want and receive targeted support from governments. And as growing body of experience reveals, this new view offers the best hope for dealing with both the old and the new agendas.
>
> (Serageldin, 1994: Foreword)

We should recall that the old and the new agenda are development and environment, respectively.

This argument is strengthened by data showing that *per capita* water production costs are four times higher in centralised than fully decentralised systems and are lowest when decentralisation is combined with centralised co-ordination (World Bank, 1994: 75). The ideal political structure for the World Bank therefore is

> a three-tiered organization, with a national agency responsible for finance, long-term planning, standard setting and technical assistance. Under the national agency, regional utilities function as operators, monitoring compliance with national standards and regulations, supervising local systems, and training local managers and technical staff. The third tier consist of local agencies that manage the local system, collect fees, monitor use and maintenance, and plan local budgets.
>
> (World Bank, 1994d: 75)

The basic function of user participation, however, seems to be to make economic and fiscal decentralisation acceptable, in particular by (1) seeking the users' consensus on the overall project and by (2) getting them to pay the increased users' fees at the local level. In other words, besides some general agreement on a given project, user participation ultimately seems to be identical to consumer behaviour, i.e. the willingness or not to pay the fees resulting from such a 'decentralised' system. Decentralisation may be effective in getting water to everybody in theory; in practice, however, users will have to pay.

Thus, decentralisation and user participation will not do away with the problem of market failure, a problem so prominent in the water sector, especially in developing countries. Subsidies may actually continue to be necessary. As the Bank admits:

> First, the amount and type of infrastructure to be supplied must be decided. Since the product will be available to all, individual choices expressed in the market cannot be relied on for decision-making. Second, investment must be made and the infrastructure must be provided. Since user charges that fully recover costs are not always feasible, private entities cannot always be relied on to make the investments. Third, infrastructure facilities must be maintained. Because many infrastructure services benefit the public at large, individuals in a market setting cannot be expected to perform this task.
>
> (World Bank, 1994d: 73)

But the Bank also says, 'Although the market clearly would fail in these functions, centralized public infrastructure bureaucracies have not proved particularly adept at performing them either' (World Bank, 1994d: 73).

Conclusion

Sustainable water resources management, as we have seen in this section, is predicated upon treating water as an economic good, privatisation and user participation, summarised under the heading of 'decentralisation'. Such decentralisation, the World Bank argues, will (1) facilitate decision-making in the water and other infrastructure sectors in favour of water projects, (2) improve investments and (3) optimise maintenance of the infrastructures once they are built. This may certainly be so, and the World Bank gives many examples where this seems to have occurred, even though we do not have any systematic evaluation of this affirmation. As such, user participation, along with economic decentralisation, may have an advantage in enticing the private sector to invest, and the public sector to take the corresponding decisions. However, such decentralisation will not do away with the ultimate challenge, which is to get the private sector to invest where it cannot see profits.

This problem is particularly acute in the case of wastewater, which is currently financed to about 30 per cent by the private sector only. This is in contradiction to water supply where the private sector currently finances 70 per cent of the investments (Serageldin, 1994: 31). Ismail Serageldin, Vice President for Environmentally Sustainable Development of the World Bank says:

> public facility projects are often characterized by a long construction period, followed by a gradual increase in the revenue extracted from the operation, with the result that the investors may have to wait 8 to 10 years before receiving their first dividend and will almost have to wait 15 to 20 years before obtaining a rate of return comparable to that offered by an industrial investment. In addition, the entire construction period may be characterized by considerable uncertainty about the ultimate profitability of the investment (because of potential cost overruns and because of the uncertainty about operating revenues). During this period of great uncertainty, remuneration of the investor's risk should compare to that of venture capital and run at the level of 25 to 30 percent. In contrast, when tariff levels are known following commencement of operation, revenues are not likely to vary as much as in an industrial project. The risk (and appropriate return) is thus less.
>
> (Serageldin, 1994: 30–1)

Another senior adviser of the Bank, John Briscoe, acknowledges that recent experiences 'show that blind advocacy of, say, water markets as "the silver bullet which will solve all problems" is not only misguided but actually counterproductive' (Briscoe, 1997: § 3.4).

Despite these considerations, the World Bank in general and Serageldin in particular seem to think that 'decentralisation', as presented above, is the key to a successful reform of the water sector. Their belief seems to be grounded in what is called the 'French model' or the 'Ruhr–French model', which has become, over time, the World Bank's model in sustainable water resources management.

The Ruhr–French model

What has become known as the 'French model' or sometimes also the 'Ruhr–French model' allows us in fact to partly reconcile the above contradiction (or, at least, tension) between treating water as a purely economic good on the one hand and the aspiration of user participation on the other. In any case it offers a way out of the problems of a purely economic approach as discussed in the previous section, and was thus adopted by the World Bank after the early 1990s. In this section we will present this model, which combines elements of participatory management with user as well as pollution charges, as it links public policy with a substantial involvement of the private sector.

The World Bank now considers that this model's underlying approach is more appropriate to the reality of developing countries, a fact which is underscored by various surveys showing, in the Philippines or in Brazil for example, the willingness of households to pay higher amounts for substantial investments that will improve the quality of lakes and rivers (Serageldin, 1994: 23). Basically, the idea of the model is that user and pollution charges should be clearly identified and linked to responsibilities and benefits at various household and collective levels. For example:

> - Households pay the bulk of the costs incurred in providing onplot facilities (bathrooms, toilets, on-lot sewerage connections).
> - The residents of a block collectively pay the additional costs incurred in collecting the wastes from individual houses and transporting them to the boundary of the block.
> - The residents of a city collectively pay the additional cost of collecting the wastes from blocks and transporting them to the boundary of the neighbourhood (or of treating neighbourhood wastes).
> - The residents of a city collectively pay the additional costs of collecting the wastes from blocks and transporting them to the boundary of the city (or of treating the city wastes).
> - The stakeholders in a river basin – cities, farmers, industries, and environmentalists – collectively assess the value of different levels of water quality within a basin, [and] decide on the quality they wish to pay for the necessary treatment and water quality management activities.
>
> (Serageldin, 1994: 20)

The structure of this model therefore embodies the central principles codified in the 1992 Dublin Statement, where the idea was that users would be involved at the lowest level, the unit of reference of managing water being the 'river basin'. In order to attain an integrated 'river basin management', the French have designed regional agencies, in total six, responsible for the general management of each river including investment, research and development (R&D), and information. There are in fact six river basin committees and six river basin financial agencies. The financial agencies are in charge of planning and macro management of the rivers. The river basin committees facilitate co-ordination among all the different parties involved in managing water resources. They fall under the joint supervision of the Ministry of the Environment (for technical matters) and under the Ministry of Finance (exclusively for financial matters). However,

> they do not enforce State legislation concerning pollution or act as contractors in the provision of services. The State laws are enforced by the 'prefets' and the contractors are essentially the private sector companies. However, the basin management agencies fulfil an advisory role in all areas of water management, and have access to considerable financial reserves which gives them real bargaining power.
>
> (Jones, 1997: 31)

Moreover, this French system also owes its popularity among policy-makers especially at the World Bank to the mix of public and private involvement in managing water utilities, deemed to result in a balance between potentially contrasted interests. Indeed, in France both water supply and sewerage is provided by private companies by means of what is generally called 'delegated management'.

But, as a matter of fact, the French modelled their system after the principles of the Ruhrverband in Germany, yet applied these principles nationwide. Box 3.3 summarises both the Ruhrverband's general idea and the French River Basin Financing Agencies.

Box 3.3 The Ruhrverband and the French River Basin Financing Agencies

Ruhrverband

The Ruhr area, with a population of about 5 million, contains the densest agglomeration of industrial and housing developments in Germany. The Ruhrverband, a self-governing public body, has managed water in the Ruhr basin for eighty years. Its 'associates' number 985 users (and polluters) of water, including communities, districts, and trade and industrial enterprises. The highest decision-making body of the Ruhrverband is the assembly of associates, which has the fundamental task of setting the budget (about $400 million annually), fixing standards, and deciding on the charges to be levied on users and polluters. The Ruhrverband itself is responsible for the 'trunk infrastructure' (the design, construction, and operation of reservoirs and waste treatment facilities), while the communities are responsible for the 'feeder infrastructure' (the collection of wastewater).

The French River Basin Financing Agencies

In the 1950s it became evident that France needed a new water resource management structure capable of successfully managing the emerging problems of water quality and quantity. Each of the country's six river basins is governed by a basin committee (also known as a 'water parliament') of between sixty and 110 persons who represent all stakeholders – national, regional, and local government industrial and agricultural interests, and citizens. The basin committee is supported by a technical and financial basin agency.

The technical tasks of the basin agency are to determine (a) how any particular level of financial resources should be spent (where treatment plants should be located, what level of treatment should be undertaken, and the like) so that environmental benefits are maximized, and (b) what level of environmental quality any particular level of financial resources can 'buy'. On the basis of this information, the basin committee determines the desired combination of costs and environmental quality for their (basin) society and how the services will be financed, relying heavily on charges levied on both users and polluters. The fundamental financial task of the basin agency is to administer the collection and distribution of the revenues.

In the French system (in contrast to the Ruhrverband) most of the resources collected are passed back to the municipalities and industries, which then invest in the agreed-on water and wastewater management facilities.

Source: Serageldin (1994: 24, Box 4).

It is also noteworthy that, in France, these agencies are not responsible for the local management of water supply and sewerage. Rather, municipalities – or intercommunal agencies, for that matter – have full responsibility for the organisation and management of water, and for this purpose, can opt between public management and the delegation of water utilities to the private sector. In general, water management in France is 'delegated' to private companies, another characteristic of the French model, i.e. so-called 'delegated management' ('gestion déléguée'). In the French case, one generally distinguishes between three different forms of delegated management, namely 'gérance', 'concession', and 'affermage'. 'Gérance' corresponds to a management agreement according to which the private contractor undertakes the obligation to perform such ancillary services as the operation and maintenance of water and sewerage facilities, provision of technical assistance, and collection of charges on behalf of the public authorities.[9] Both 'concession' and 'affermage' agreements provide for the obligation of the contractor, respectively the 'concessionnaire' and the 'fermier', to operate local utilities at his own risk, while the municipality retains full ownership of the existing infrastructure and the works performed in the life of the contract (Finger and Lobina, 1999: 40). The difference between 'concession' and 'affermage' resides in the fact that the investments needed for the infrastructure fall under the responsibility of the 'concessionnaire', while it is the municipality that is responsible for these investment in the case of 'affermage'.[10] It appears that in practice, affermage is the prevailing arrangement in France today. In World Bank terminology (see p. 82), 'concession' would be close, yet not identical, to the Bank's idea of a concession or BOT, 'affermage' to a lease contract, and 'gérance' to a management contract. However, it is clear that the French approach is more elaborate and also more rooted in practical experience, whereas the World Bank's categories are above all theoretical in nature.

It is thus not surprising that, under the influence of the French model – of both river basin and delegated management – the World Bank has somewhat adapted its conceptual framework and established its new policy in water resources management, thus overcoming the shortcomings of its original purely economic approach presented above. As we will show in Chapter 4, it is also this French model which gives rise to (mainly French) water transnational corporations (TNCs). These TNCs are then, a result of the fact that the World Bank adopted this French model, becoming particularly successful worldwide. But let us first examine how this French model is being put into practice by the World Bank.

9 In the original conception, gérance implies the contracting out of the whole operation and can also encompass elements of profit-sharing between the employer and the contractor; see Roth (1987: 251–6).

10 Rossi (1997: 162–4); Roth (1987: 251–6).

Follow the model

The World Bank's new model in water resources management, as outlined above, was put into practice after 1995. The first experience in reforming the water sector took place in the Kingdom of Morocco, leading to the first conclusions and recommendations issued by the Natural Resources and Environment Division from the Maghreb and Iran Department of the World Bank. Our second example is Yemen. Although both countries are from the Middle East and the North Africa region, their water resources problems are not at all identical. Yet, interestingly the recommendations the World Bank proposed are very similar. Our third example, Bangladesh, demonstrates how the World Bank systematically applies its general model notwithstanding the particularities of each country.

Morocco

As said above, Morocco, for the World Bank, was to be a model in water resources management and water sector reform more generally: 'in implementing its broad commitment to a comprehensive approach to water resource management, Morocco could provide a model from which others may learn' (World Bank, 1995a, Executive Summary, § 1 & 2).

In Morocco the challenges were seen by the World Bank as follows:

1 The decline in *per capita* supplies and exhaustion of available resources in a number of key basins;
2 The degradation of water quality, which has reached critical levels in the primary basin;
3 The provision of only 14 percent of the rural population with potable supplies;
4 The rapidly rising marginal cost of water and sharp increases in investments costs, which are not matched by cost recovery;
5 The inadequate demand management, as reflected in losses of over sixty percent of irrigation water and twenty-five percent of urban potable water supplies;
6 The expected thirty percent decline by 2020 in water available per hectare, which, together with higher water prices and lower output prices, requires sharp improvements in farm productivity and water efficiency;
7 The relative absence of efficiency and equity criteria in allocation of increasingly scarce water;
8 The insufficient stakeholder participation in water resources management; and
9 The need for reform of the legal and regulatory system to address the challenges identified above.

(World Bank, 1995a: Executive Summary, § 4)

There is indeed first the challenge of declining groundwater stocks in most of Morocco. The situation is particularly threatening in the Tensift and the Souss basins. In this case, the problems stem from uncontrolled groundwater mining. Underlying that, there is a competition between the coastal tourist industry around Agadir and the farmers in the Souss basin. Secondly, pollution has caused significant degradation in water quality, and in some cases potable water supplies have had to be suspended in the Sebou basin owing to excessive pollution loads. The threat here stems from the fact that, according to World Bank estimates, virtually all renewable water resources will be mobilised to match demand at the national level. Moreover, at present, wastewater reuse for potable consumption is not envisaged, as the health risks and costs of adequate treatment are considered prohibitive. (World Bank, 1995a: 12). The third challenge reflects the main issue in water resources management, i.e. development versus environment. As the World Bank says:

> Over half the population lives in rural areas and barely 14 percent of these citizens have access to safe and reliable supplies of water. The virtual absence of potable water supplies is considered the primary cause of rural ill-health. Rural families on average devote one and a half hours a day fetching water and pay up vendors up to a hundred times the urban fee charged by the drinking water utilities. The absence of potable supplies also has retarded a broad spectrum of rural commercial activities. The provision of potable water therefore offers a means to reduce rural poverty. The implementation of this objective is a key challenge.
>
> (World Bank, 1995a: Executive Summary, § 7)

The problem therefore is to reconcile both objectives, environment and development, knowing that Morocco's current water consumption pattern will bring the country in the very near future, and according to international categorisations, from a 'water-stressed situation' to a 'chronically water-stressed situation'. Indeed, according to World Bank figures, in 1995, the *per capita* renewable water resources were estimated at $800m^3$ whereas they would decline to $400m^3$ by 2020, if the Kingdom of Morocco followed the same policies. (World Bank, 1995a: 1). Therefore, the need for reform is essential, despite the fact that this argument is valid for most of the countries in the world. The fourth to the ninth challenges pertain mostly to recommendations proposed by the World Bank in order to change the current pattern of consumption, while ensuring long-term development. The fourth challenge particularly clearly indicates the desire to consider water as an economic good when saying:

> Masking the scarcity of water delays the inevitable, which is a revision of tariffs across sectors to reflect the true economic value of water. To the extent that the projections may overestimate future demand, they may

> also provide a justification for the acceleration of dam construction and other investments. Given the uncertainties associated with projections, the principal objectives should be to ensure that consideration is given to the true economic cost of water and its likely evolution. The introduction of pricing considerations may not only encourage water conservation, but also to foster improvements in technology to extract new and non-conventional resources.
>
> (World Bank, 1995a: 17)

In its new vision of water resources management, the World Bank's victims are clearly the rural areas and irrigation, as the main goal is to transfer water from the development of irrigated area to the urban area.[11] This is explicitly stated in the report when it is said that the best way to improve water efficiency and to curtail water use in irrigation is to reduce the irrigated areas. Although the World Bank recognises that such measures are extreme, the report hints that, in the longer term, a reduction in the irrigated areas may prove to be necessary (World Bank, 1995a: 25–6). According to the logic of the World Bank, privatisation of this sector is therefore the best option (World Bank, 1995a: 25–6), since it will theoretically reduce the costs while at the same time limit the pressure locals could have on a privatised rather than on a state-owned company. Moreover, quality norms will be used so as to justify the increase in the price of water. As the Bank says:

> The combination of stricter environmental regulations and higher water charges over time is expected to increase efficiency in urban and industrial demand and this is reflected in the base demand calculations. Efficiency gains also can be achieved through more effective billing procedures, decreased leakage in the infrastructure, and better monitoring of fraud control measures. Efficiency of delivery, measured as percentage of water entering the municipal system that is actually billed to customers, is estimated to be seventy percent. This is projected to rise to eighty percent by the year 2020 reflecting both a reduction in water losses and an improvement in cost recovery. In urban areas, as well as in agriculture, it

11 For example, the report takes the example of Souss:

> In the Souss, the available water supplies have already been stretched beyond the sustainable limit. Growing pressure from urban and industrial development has led to mining of groundwater. A reduction in irrigation demand would provide a sustainable solution and could be achieved by improving agricultural efficiency or reducing irrigation perimeters. The establishment of compensatory arrangements or tradable water rights *could facilitate a transfer of water from irrigation*. Urban, industrial or higher value agricultural producers could compensate farmers with a lower return to water for reductions in their water use, allowing the farmers to invest in water-saving technologies or in diversifying their activities.
>
> (World Bank, 1995a: Executive Summary, § 13)

> is clear from the above that great scope exists for demand management. Increasingly, this is expected to become a more prominent feature of water resource management in Morocco.
>
> (World Bank, 1995a: 26–7)

In short, the Moroccan water situation is critical. In a few years, most of the water resources will be used, while wastewater treatment is not considered owing to its high cost. The proposed solution, according to the World Bank, is to introduce the concept of water as an economic good as a priority and, by doing so, transfer water from the irrigated to the urban areas. The subsequent solution is to gradually increase the prices according to demand. The Bank says that these two actions can (and should) be carried only by means of privatisation, since privatisation will somewhat limit local pressures, which was not the case with the state-owned company. As part of the general reform of the water sector, the World Bank also emphasises the need for decentralisation.

Indeed, the last two challenges are related to institutional issues. Consequently, and entirely in line with its philosophy, the World Bank emphasises the need for decentralisation, as well as greater participation of stakeholders. In fact, 'decentralisation' here means the disengagement of state activities motivated by the willingness to lower the budget deficit, while handing local networks over to private actors. As the Bank says:

> Public expenditures in Morocco are strictly circumscribed by the commitment of the government not to extend its deficit budget. Indeed, the government intends to gradually reduce its deficit and to record a surplus by the end of this century. Rapid growth is projected to facilitate a moderate rise in spending. Nevertheless, without a major reallocation of government expenditure, away from the military or social ministries, which seems unlikely, expenditures on the water sector may at best be expected to rise at the pace of revenue raising. Under these circumstances, alternatives sources of revenue for as much as four billion DH [Dirams] of water sector investments per year will need to be found. Greater financial autonomy, extending where appropriate to privatization, for parts of the water sector provides a possible mean s to achieve the ambitious investment targets without compromising the financial integrity of the government.
>
> (World Bank, 1995a: 21–2; 40 million DH is equivalent to $3.7 million (or 4.1 million Euro) according to the exchange rate at 20 April 2001)

Having borrowed from different ideological currents and models, the World Bank thus ends up with some serious contradictions when it comes to water resources management. Let us mention the two main such contradictions:

- There is indeed no clear definition of what is meant by 'decentralisation'. Participation at the local level cannot be correctly applied if the underlying reason is to lower the national budget deficit, as the local actors have even less capacity to solve such financial problems, and decentralisation thus simply means the opening of the market to the private sector. In other words, decentralisation will more or less automatically end up with privatisation.
- The majority of the remedies recommended by the World Bank do in fact not lead to decentralisation and local participation at all. Rather, they emphasise the need for a better national co-ordination and a strategy of river basin management along the French model. The problem here stems from the fact that the French model is a quite complex model, linking a centralised state with a decentralised water system.

And this French model is clearly a contradiction in the World Bank's recipes, in particular privatisation. After all, river basin management derives from a centralised model, after which Morocco's near-term strategy was actually modelled. As the Bank says: 'Morocco's near term strategy includes smoothing water balances across basins, slowing depletion of non-renewable groundwater resources, and improving overall efficiency of water infrastructure by strengthening water networks within basin' (World Bank, 1995a: 19). But in reality, this is not what is happening. Let us again quote the Bank:

> By the year 2000, urban water tariffs are expected to provide for full cost recovery of the urban water distribution systems, including debt repayment, as well as the operation and maintenance costs of the sewerage systems. Once this target has been reached, it is anticipated that the next step would be for urban tariffs to cover the costs associated with the mobilization of urban water supplies (dams and transfers) as well as the costs of investment in sewerage systems. The improvement in provision of services and increased managerial and financial efficiency within the urban water supply sector within the next decade could be expected to be associated with a process of privatization. Drawing on a comparative experiences in other countries, much of the urban water supply distribution sector may already be in the position to enter a phase of privatization based on competitive tendering of fixed term lease contracts or concessions.
>
> (World Bank, 1995a: 32)

In other words, privatisation will take place in the most successful sectors where cost recovery is possible, which will leave the State with the task of developing investment projects that will eventually be profitable in the long run. Privatisation might thus mainly occurs in the urban areas, while the countryside and agriculture will thus most likely remain in the hands of the State. Yet, successful river basin management will have to include both cities and countryside. By combining neo-classical economic principles with a state-centric French model, the Bank may well have ended up with an insurmountable contradiction. In other words 'sustainable water resources management' as conceived by the Bank might be a fundamentally contradictory concept.

Yemen and Bangladesh

Yemen and Bangladesh are two countries which do not share the same problems in water resources management. While Yemen, despite being among the oldest irrigation civilisations, faces one of the most worrying water crisis in the world,[12] Bangladesh is confronted with seasonal problems, i.e. with both a dry and a wet season characterised by floods. Nevertheless, besides the different challenges in both countries, many similarities can be found in the solutions proposed by the World Bank. Before looking at both situations, it is worthwhile noting that historical structures of the countries are rarely taken into account in the solutions proposed by the World Bank. Indeed, there is a considerable difference in recommending the French model of river basin management to a country that has a federal or a centralised system, for example. Also one must note that the cultural and religious significance of water is never even mentioned in the World Bank reports.

Yemen and Bangladesh are indeed very specific cases, barely comparable with most other countries. In particular, their water problems have little in common with our previous example. We want to show here that the World Bank is basically applying the same model to all countries. Yet, in all cases we discover the fundamental contradiction rooted in the very concept of water resources management.

Bangladesh's basic problems stems from the difficulties of dealing with different water management approaches needed for wet and for dry seasons. As emphasised in the corresponding report,

> in the wet months, when flooding is common, water management entails flood protection, proofing, and drainage. The task is complicated by the varied impact of floodwater on different economic activities. For example, agriculture requires flood protection and speedy drainage of water to avoid crop damage, while fisheries and navigation depend on floodwater for support. Such circumstances naturally breed conflicts. In

12 The World Bank forecast that if no change were made in water resources management, the country would become simply a desert. See World Bank, *Yemen, Towards a Water Strategy: An Agenda for Action*, Washington DC: World Bank, Report No. 15718-YEM (13 August 1997), hereafter World Bank (1997d).

> the dry season, water management takes a different turn. There must be enough water for drinking as for municipal, industrial, agricultural, and environmental use. In formulating a water policy, Bangladesh will have to establish priorities for all these different uses, designing a holistic strategy that considers environmental factors while juggling the need of various sectors.
>
> (World Bank Bangladesh, 1998: Executive Summary)

The problems in Yemen are of a completely different nature. Like most other Middle Eastern countries, population growth and market-led agriculture are the two main explanations for the increasing water stress. More precisely, the World Bank says, '(1) groundwater is being mined at such a rate that large parts of the rural economy could disappear within a generation, (2) major cities have grown very short of water, and (3) many people, particularly the poor in the cities and the countryside, do not have access to safe water' (World Bank, 1997d: 2). Yemen thus somewhat resembles the Moroccan case, though the seriousness of the situation in Yemen should lead to stricter measures.

Having underlined the different problems in both cases, what appears as striking is the way in which both countries will have the same model imposed on them, which seems to fit whatever the difficulties encountered in the water sector. In both cases, the World Bank suggests, first, to treat water as an economic good, secondly the need for decentralisation and user participation and, thirdly, a push towards privatisation. The most controversial aspect is, again, the Bank's willingness to decentralise, which will lead to privatisation, leaving the State with everything that cannot be privatised. Indeed, decentralisation can hardly lead to anything other than privatisation, given the fact that, in the case of Yemen and elsewhere, the Bank never says anything about how to develop institutional measures of community involvement. The explanation, of course, lies in the fact that the World Bank considers public institutions in particular and the State in general not to be efficient. Public institutions are said to have too many staff, while state policy is based on water resources development, rather than sustainable management. In addition, the Bank says in the case of Yemen, public programmes have tended to reinforce an inequitable distribution of water:

> one example is targeted and subsidized agricultural credit through CACB (Cooperative and Agricultural Credit Bank). In effect, it is the better off who generally succeed in getting hold of this cheap money, and then often failing to repay . . . public policy has given the landowners access to the cheap credit and equipment that enable them to drill. The poor who lack the means to develop their own water resource then have to pay for irrigation or rely on rainfed cultivation. Finally, potable water supply in towns has concentrated on established areas rather than on the poorer

> neighbourhoods, and in rural areas on those governorates nearest to the capital, predominantly around Sa'ana.
>
> (World Bank, 1997d: 6)

Yet, Yemen's problems, according to the World Bank, primarily stem from the fact that there is a clear lack of government structure in the water sector (World Bank, 1997d: Introduction).

Thus, the macro solution proposed by the Bank is to move towards efficiency prices for water, when it says:

> government is able to send powerful signals to users through the pricing system. The challenge is to remove all the distortions and incentives that have led to overpumping of groundwater – effectively to change relative prices to discourage groundwater use. The agenda might include: introducing a higher price for diesel, levying higher tariffs and taxes on pumping equipment, eliminating credit subsidies – and perhaps eliminating credit programs – for pumps, and removing current incentives (such as import controls) to the production of agricultural products that are water intensive, notably qat. Making these changes would carry significant costs of adjustment, some of which would be felt in other sectors (e.g., the transport sector would also pay the higher diesel price). However, action on relative prices has the strong merit that government could actually do it, whereas other actions like trying to regulate and control water user behaviour at the well head are much more difficult in the Yemeni context. The adjustments would have to be phased to allow users to reposition themselves in the economy.
>
> (World Bank, 1997d: 11)

However, the fundamental question remains: how to reconcile the real value of water, privatisation, and serving the poor.

In Bangladesh, the same line of thought was followed, despite the fact that the problems are completely different:

> Policies for integrated water resource management should support sustainable development and protect investments in the short and long term. Such policies would include assigning common property rights in floodplains, holding users and developers responsible for water quality, sharing costs between users and developers, and ensuring that public and private bodies share responsibility for maintenance of water systems.
>
> (World Bank, 1997d: Executive Summary)

The State and its institutions are, not surprisingly, again seen as the main ones responsible for the mismanagement of water resources:

> First, the public institutions, virtually unchanged since their founding a few decades ago, are unable to meet the demands of present-day water resource management. Second, they have become much too heavily staffed, impeding the delivery of public services. Third, their original top-down hierarchy of command requires change. Fourth, their relationship with the public is autocratic rather than democratic. And finally, they have not given private-sector enterprises enough room to grow and take over many of their functions.
>
> (World Bank Bangladesh, 1998: Chapter 1)

Behind this admittedly ideological critique, decentralisation appears, again, as the most promising instrument for better water resources management. 'The state's participation in water management should be limited to essential acts supporting the creation of public goods and infrastructure and dealing with market failures . . . The success of these institutions will depend heavily on grassroots water management organizations – whether they are efficient, share responsibility, and respond innovatively to the changing needs of a wide variety of water consumers' (World Bank Bangladesh, 1998: Executive Summary). In this regard, the World Bank suggests that that the most important goal of institutional development in the water sector is to make the system responsive to peoples' needs, adding that 'redesigning institutions therefore requires public participation at all different levels. Water organizations are either public or private agencies, depending on their roles and responsibilities' (World Bank Bangladesh, 1998: Chapter 4). Yet again, we find that decentralisation appears to be very closely linked to privatisation:

> there are a number of private firms in Bangladesh capable of providing various services to the water sector such as design, construction, and operations and maintenance. In light of the government's willingness to encourage the private sector to take over functions from state agencies, it would behoove the sector to have well conceived policies in place for private operation. Major public organizations have awarded operational contracts to private companies for agricultural services in the past, and the same thing could happen in the push to decentralize water services.
>
> (World Bank Bangladesh, 1998: Chapter 4)

As part of the proposed reform, the World Bank emphasises, in Bangladesh and elsewhere, the need for greater user participation, when saying:

> it is critical to develop organizations at the field that manage the system through consensus. One possibility is to empower local government to form water management committees with representation from each user groups at fixed intervals. The local government body should be an integral

> part of the process, with the power of adjudication during disputes. Local committees could be integrated into than a water management committees, serving as local representatives in national efforts to plan and develop water projects.
>
> (World Bank Bangladesh, 1998: Chapter 4)

While such user participation is certainly desirable from our point of view, the question remains as to whether a private firm would be inclined to work under such a structure.

Besides decentralisation, the other noteworthy aspect of the World Bank's policy is its attempt to try to convince governments to consider water as an economic good:

> The job of managing water resources is problematic because of the controversy over allocation rights and the difficulty of enforcing rules and regulations on conservation. If water were treated purely as an economic commodity, consumption would be controlled through pricing. But water is a public good subject to non-economic criteria for allocation. Different countries have struggled with the problem of managing demand using both price and non-price mechanisms. Their experience provides useful insights for Bangladesh.
>
> (World Bank Bangladesh, 1998: Chapter 4)

To consider water as an economic good implies of course dramatic consequences for irrigation. 'For efficiency's sake, it would help to include at least some aspects of a market-based system, such as allowing growth of water markets for a particular use like irrigation' (World Bank Bangladesh, 1998: Chapter 4). In Bangladesh, as in most of South Asia, the development of informal local water markets in the irrigation sector is already a reality and according to the World Bank, the development of this market needs to be promoted despite some concerns:

> The main concern about developing a private sector water market was that irrigation equipment would be concentrated in the hands of higher-incomes groups, which might lead to differential pricing and inequitable access to irrigation. The Bangladesh Institute of Development Studies developed two large-scale sample surveys covering the same households; these indicate that irrigation increased substantially between 1987 and 1994, more for the relatively large farmers than for the medium and small ones. The difference in access to irrigation between small and large farmers remained. The important point is that during this period of large-scale private market expansion, the irrigation charge paid by the farmers did not increase. In fact, the average water charge declined by 4 percent, while the price of rice increased by nearly 30 percent. Thus, in real terms,

> irrigation water has become substantially cheaper after liberalization of the water market in Bangladesh.
>
> (World Bank Bangladesh, 1998: Chapter 4)

We can see therefore that in spite of the fundamental differences of the three case studies, a magical solution has been applied to all: decentralisation, and in particular privatisation, constitute one aspect of the reform, while considering water as an economic good constitutes the other. Participation, if not simply used by the World Bank as a slogan to make privatisation more acceptable, is in fact contradictory to the very principle of considering water as an economic good, as it introduces non-economic considerations, and thus distorts the market.

Conclusion

In this chapter we have examined the two main policies of the World Bank, one which is concerned with reforms in infrastructures in general, and which can be linked to the processes of deregulation and privatisation as presented in Chapter 1. The other policy pertains to environmental concerns in general and water concerns in particular, and seeks to address the issue of water stress and therefore refers to the Dublin principles (e.g. water as an economic good). From these two policies combined emerged the concept of (sustainable) water resources management, which, some time during the 1990s, was put into the larger framework of the 'French model'.

In infrastructure, management problems in the public sector, along with neo-liberal considerations on the role of the State, were the main reasons for reform. Such management problems were, according to the Bank, characterised by the misallocation of infrastructure investments, relying too much on new investments and not enough on maintenance, but also by the lack of competition and responsibility. This criticism lead the Bank to conclude that privatisation of the infrastructure constituted the best way to eliminate underpricing, to improve technical efficiency, as well to increase quality of service by means of competition. The first measure implies that water be treated as an economic good, with the immediate effect that this will lead to higher prices for consumers. The second implies that efficiency gains can be made and redirected, for example, towards the billion people without water. The third measure implies that market principles (e.g. competition) will replace monopoly structures. We have criticised the underlying assumptions of these measures, namely the idea that private companies will proceed to 'cream-skimming', thus tend to do whatever is profitable, while leaving the state with all the unprofitable aspects of water supply and sewerage. We have also highlighted the fact that water will remain a natural monopoly, which by definition limits the possibilities for a competitive market.

In environmental matters, as we already underlined in Chapter 2, the pressures and challenges in the water sector are tremendous and immediate reforms are needed in order to move towards a more sustainable development path. The most obvious response to such pressures and challenges for an economist is to define the real price of water according to its scarcity so as the limit the pressures on water resources. This consideration brought, as we have seen, most specialists to consider water as an economic good. The main problem here is to put such economic principles into practice, not to mention the difficulties in elaborating the tools and data needed. We have also stressed in this chapter that putting such ideas into practice will lead to exorbitant prices, especially in the agricultural sector.

Yet, the link between infrastructure development policies and environmental concerns requires the application of such economic principles. At this purely economic level, both policies are therefore compatible. However, this link does not necessarily lead to privatisation (or in Bank terms 'market decentralisation'), as the World Bank seems to imply. That the World Bank nevertheless ends up with the idea that privatisation is the answer to water resources management problems can be explained by the fact that considerations pertaining to infrastructure development ultimately seem to be more important in the Bank than environmental (and social) considerations. This tension, or even contradiction, in setting priorities is even acknowledged by senior advisors of the Bank, who clearly state that the problem of financing cannot be overcome by private sector participation only. Decentralisation therefore appears to become the solution to this environmental concern, conceptualised as it is around the French model of river basin management.

It is by combining these two somewhat contradictory trends that the Bank has come up with the concept of water resources management. This concept has then been applied as the magical solution, regardless of the problems of each country. In all three examples we have briefly presented here – Morocco, Yemen, and Bangladesh – we have been able to identify the World Bank's underlying model, which is (1) to treat water as an economic good, (2) to push for decentralisation and (3) to encourage privatisation. Yet, in all three countries two main problems remain unanswered in the different reports of the World Bank we have examined. The first problem is how to reconcile the real value of water, the process of privatisation, sustainable development, and serving the poor.[13] This problem is becoming increasingly urgent, and several World Bank programmes have been launched recently in order to reconcile all four objectives, such as the Business Partners for Development programme. The second problem pertains to the 'decentralisation versus centralisation'

13 The World Bank is now concentrating on this problem. For example, between 31 May and 2 June 2000, an international workshop, entitled 'Infrastructure for Development: Private Solutions and the Poor', co-sponsored by the World Bank and the UK government, was examining the relationship between private sector participation and the access to the urban poor. See also Brook Cowen and Tynan (1999).

dilemma. Indeed, the World Bank is clearly pushing towards decentralisation. Yet, such decentralisation simultaneously requires a national organ responsible for inter-sectoral water resources management, as well as a regulatory framework. This dual process is not very clearly thought through in the various World Bank documents, given the fact that the Bank adheres to a quite simplistic view of privatisation. We will address this second problem in Chapter 7. Most specialists nowadays, in particular those of the World Bank, seem to be searching for new solutions in water resources management and many problems remain unanswered. For example, let us refer to some research conducted by the World Bank, which shows that reforms in water pricing have, in most of the countries studied, fallen short of their original objectives. As explained by Ariel Dinar, many of the shortcomings are due to the economists' and water experts' failure to include political considerations in their economic approach (Dinar, 2000: 5–6). The integrated approach recommended by the Dublin principles therefore still has a long way to go before World Bank specialists recognise the limits of economic theory and modelling.

In Chapter 4, we will examine how this conceptual weakness and underlying preference for privatisation by the World Bank have actively promoted the growth of TNCs in the global water sector. By doing, so, we will argue, the World Bank has now created an additional problem in water resources management, a problem, which it will be very difficult to tackle in the future.

Chapter 4

The emergence of public services TNCs and their strategies in the water sector

> [We need] an increase in private sector investments in water infrastructure from the current levels of $16–20 billion to about $125 billions per year.
>
> (Cosgrove and Rijsberman, 1999)

This chapter will focus on one of the key factors currently affecting the international water governance system, without which the actual profound transformations in the water sector cannot be fully understood. This is the recent rise and rapid expansion of transnational corporations (TNCs) in general and in the water sector in particular. While we are especially interested in the TNCs operating in the water sector, it is obvious that their evolution cannot be separated from the larger dynamics of globalisation, which has made the rapid growth and expansion of TNCs possible. In the first section we will therefore discuss the evolution of TNCs in relationship to economic globalisation, with a particular look at concentration among these TNCs. This will give us the general background against which the evolution of water TNCs must be seen.

But TNCs are not limited to purely commercial activities. Indeed, one can also observe a growing number of TNCs, which make their business in the public sector, in particular by means of the type of partnerships described in Chapter 2. Not surprisingly, some TNCs active in the larger infrastructure areas – e.g. energy, transport, telecommunications, water, etc. – count today among the biggest TNCs worldwide, as we shall show. Moreover, one can say the role of the World Bank in financing such infrastructure developments in the developing countries has been key in the growth of these 'infrastructure TNCs'. TNCs, which make their business by providing public services in the various infrastructure areas are often also called 'public services TNCs' (e.g. Hall, 1998). A rapid look at the 2000 Fortune 500 companies shows that twenty-six firms within the first 200 are in fact public services TNCs (i.e. TNCs, which make the bulk of their money from providing public services) (see Table 1.1, p. 16).

In the second section of this chapter, we shall in particular examine the main public service TNCs active in the water sector, which are mostly of French origin. These are Vivendi, Suez Lyonnaise des Eaux, and Bouygues/SAUR. We will see how these three TNCs have grown parallel to economic globalisation but also thanks to the World Bank's new models of public–private partnerships (PPPs). We will also see how these French public services TNCs have had a significant historical advantage over other more recent companies seeking to move into the global water sector and beyond, given their privileged links with French politics, but also given the fact that the Ruhr–French model was adopted by the World Bank as the model of sustainable water resources management (Chapter 3). Building on these advantages and their technical expertise in the water sector, they have developed a strategy of expansion throughout the public services, and have become among the biggest, if not the biggest, public service TNCs today. We will also present and discuss in this section the rise and evolution of British water companies, whose emergence directly stems from privatisation in the United Kingdom in the late 1980s. Given the fact that the European Union (EU) is also planning to liberalise the water sector in the very near future, TNCs from other European countries are also moving into the water sector and will be briefly discussed here. Yet, after a brief period of TNC diversity, we can now also observe the same trend of concentration in the water sector, which makes it likely that some three or four public services TNCs will dominate the world water market in the near future.

The last section will synthesise the different trends in the evolution of TNCs in the water sector. We shall see how TNCs have entered the water sector, encouraged as they were by the World Bank water policy, at least in so far as developing countries are concerned. We will also try to crystallise these water TNCs' overall expansion strategies, first in developing but increasingly also in developed countries. This will lead us finally, to discuss such concepts as 'multi-utilities' or 'multi-service', which is certainly the most apparent strategic behaviour among water and other public service TNCs today.

Globalisation and the growth of TNCs

This section follows directly from Chapter 1 on globalisation. Indeed, in that chapter, we saw how economic globalisation, growing trade and cross-border investments have favoured the emergence of TNCs. This process, as we saw, considerably accelerated since the late 1980s, and as such was not unrelated to the end of the Cold War. Depending on how one defines a TNC, their growth is indeed astounding. In the 1950s, 'TNC' was not even recognised as a term. At best, it was just a formal unit of analysis (e.g. Dunning, 1998). It is only after the mid 1970s that foreign direct investment (FDI) was increasingly seen as a firm-driven managerial decision, rather than as a capital inflow into a country (Hymer, 1976). With the emergence of these TNCs, much of the literature and conferences then turned to public policy and to the regulation of TNCs. Indeed, already in the early 1970s, studies were conducted as to the impact of TNCs on development and international relations (UN, 1974). Basically, the question was how to control this seemingly uncontrollable phenomenon. In the late 1970s, the International Labor Organization (ILO), the United Nations Center on Transnational Corporations (UNCTC) and the Organisation for Economic Cooperation and Development (OECD) all tried to design codes of conduct for TNCs. This appeared to be a difficult task, given the fact that TNCs were rapidly exploding.

Indeed, in the beginning of the early 1990s, the number of TNCs was evaluated at around 37,000, controlling some 170,000 foreign affiliates (UNCTAD, 1993: 19). It is estimated that there are now approximately 63,000 TNCs, with 700,000 foreign affiliates (UNCTAD, 2000: 9). It seems that there does not really exist a counter-power to these global corporations, leading leaders such as the UN Secretary General Kofi Annan to appeal to TNCs to cast universal principles in the areas of human rights, worker's rights and the environment in a *voluntary* manner. TNCs, even in the eyes of world leaders, seem to be an irreversible and uncontrollable phenomenon.

Also, there seems to be a general agreement as to the definition of a TNC. TNCs basically are companies which own or control production, services, or facilities in at least two countries. For example, Raymond Vernon defines a TNC as 'a network of related enterprises, composed of a parent in one country and subsidiaries or affiliates in other countries'(Vernon, 1992: n. 1). The UN consider that

> in the broadest sense, any corporations with one or more foreign branches or affiliates engaged in any of the activities mentioned may qualify as multinational. More strictly, a particular type of activity (e.g. production), a minimum number of foreign affiliates (e.g. six), or a minimum foreign share of activity (e.g. 25 per cent of sales or assets) may be added as conditions for qualifying for the definition.
>
> (UN, 1973: 5)

Some authors, such as Korten, an analyst of TNC activities and former Harvard professor, clearly distinguish between multinationals and transnationals. For Korten,

> a multinational corporation takes on many national identities, maintaining relatively autonomous production and sales facilities in individual countries, establishing local roots and presenting itself in each locality as a good local citizen. Its globalized operations are linked to one another but are deeply integrated into the individual local economies in which they operate, and they do function to some extent as local citizens. The trend, however, is toward transnationalism, which involves the

> integration of a firm's global operations around vertically integrated supplier networks . . . Although a transnational corporation may choose to claim local citizenship when that posture suits its purpose, local commitments are temporary, and it actively attempts to eliminate considerations of nationality in its effort to maximize the economies that centralized global procurement makes possible.
>
> (Korten, 1996: 125)

However, it seems that nowadays the terms TNC and multinational corporation (MNC) have become more and more interchangeable. For example, the definition given by Joan Spero and Jeffrey Hart (1997) for an MNC, i.e. 'an entreprise that engages in Foreign Direct Investment (FDI) and that owns or controls value-added activities in more than one country', is also applicable, in their own words, for multinational enterprises (MNEs), transnational corporations (TNCs), and transnational enterprises (TNEs) (Spero and Hart, 1997: 96). Nevertheless, all authors using these terms agree that transnationalisation increasingly constitutes the prevailing model of corporate structure. Indeed, they say, 'it is only in our lifetime that the role of trade, foreign direct investment (FDI) and cross-border strategic alliances have become such critical determinants of economic progress. And, it has only been in the last twenty years that the world economy has become so structurally interdependent that the use of the word *global*, as distinct from *international*, has become justifiable' (Dunning and Hamdani, 1997: 12).

In the literature, the growth of TNCs appears to be very closely related to a series of factors, in particular trade and FDI. As shown by Spero and Hart, there have been a number of theories explaining the rise and strength of TNCs, the most popular one being the so-called *internalisation theory* (e.g. Buckley and Casson, 1976; Rugman, 1996). Dunning has developed what he calls an eclectic theory, which includes internationalisation, but explains TNC growth by the combination of three different advantages, (1) ownership, i.e. specific advantages of property rights and intangible assets, (2) internalisation incentives advantages to overcome market failures such as buyer uncertainty and the lack of future markets and (3) location-specific advantages, which include differences in country natural resource endowments, transport costs, cultural factors and government regulations (Dunning, 1980).

Another theory explaining the rise of TNCs is the so-called *product cycle theory*, developed essentially by Raymond Vernon (1996).

> The product cycle theory argues that firms expand abroad when their principal products become 'mature' in domestic markets. During the initial or rapid growth stage of product commercialization the firm attempts mainly to respond to domestic demand. As growth tapers off, the firm may begin to look for new sources of demand in export markets. Eventually, domestic demands begins to fall as the market is saturated,

> new firms begin to challenge the earlier entrants to the markets, and the firms looks for ways to protect its revenues and profits by establishing foreign subsidiaries with lower factor costs so as to remain competitive in the home market and/or better market access to foreign markets.
>
> (Spero and Hart, 1997: 110)

Along the same line of arguments, the obsolescing bargain theory explains the advantage of firms in the short–mid-term to expand abroad thanks to a good bargaining position of TNCs with host countries' governments because of firm-specific advantages, such as superior technology, access to capital markets and access to final product markets. Another major theory which has been developed by Stephen Hymer (1976) is the oligopoly theory of foreign investment, which contends that firms move abroad to exploit the monopoly power they possess through such factors as unique products, marketing expertise, control of technology and managerial skills, or access to capital. Finally, another major theory of TNCs and FDI flows is the tariff-jumping hypothesis, which explains some attempts by TNCs to jump over tariff or non-tariff barriers (NTBs) by establishing foreign subsidiaries (Lunn, 1980; Culem, 1988).

Much of such internationalisation or transnationalisation is achieved not by means of expansion, but by means of strategic alliances with already-existing companies, mergers and acquisitions. Box 4.1 lists the main reasons for such cross-border strategic alliances.

Liberalisation and privatisation have also been an important factor in mergers and acquisitions (M&As). For example, cross-border M&As into Central and Eastern Europe doubled to $10 billion in 1999, especially in Poland and in the Czech Republic (UNCTAD, 2000). The European Union (EU) is also illustrative of this process. Indeed, European FDI developments in 1999 were more than ever driven by M&As. The value of cross-border M&A-related sales and purchases increased by 83 per cent and 75 per cent, respectively, reaching $345 and $498 billion. As a result of liberalisation, the EU accounted for almost half of all global cross-border M&A-related sales and 70 per cent of purchase, and EU companies were involved in all but one of the ten largest cross-border M&As last year (UNCTAD, 2000).

There are, indeed, many very good reasons why to go beyond one's national borders. Not surprisingly, therefore, cross-border M&As abound, and the amounts involved grow higher and higher. In 1990, for example, the value of global M&As amounted to approximately 0.5 trillion $US, but ten years later the number had increased sevenfold to 3.5 trillion $US. Table 4.1 lists the biggest M&As of the year 1998.

It appears clearly that most recent mergers have mainly occurred in the financial and the telecommunications industries. But mergers take place in other sectors as well, leading to growing sectoral concentration. New actors, as illustrated in Tables 4.2 and 4.3, are now part of the largest TNCs.

Box 4.1 Reasons for cross-border strategic business alliances (Dunning, 1993b: 250)

- To capture economies of synergy (e.g. by pooling of resources and capabilities, and by rationalisation of production).
- To lower capital investment, to disperse or reduce fixed costs, to better exploit scale and/or scope economies, to lower unit costs by using comparative production advantages of each partner.
- As a consequence of the convergence of technologies and interdependencies among innovation processes, to spread R&D costs, to gain speedy access to new technologies.
- As a response by firms to growing competition, a shorter product cycle and a faster rate of technological obsolescence.
- To obtain reciprocal benefits from the combined use of complementary assess, to exchange patents and territories.
- To overcome government-mandated trade or investment barriers.
- As a means of promoting joint R&D and design efforts with suppliers and/or customers.
- To assist the entry process of small firms into high-risk, entrepreneurial ventures, especially in emerging technology sectors.
- To gain new knowledge about, or achieve quicker access to, markets and/or to spread marketing and distribution costs, to widen market sources.
- To pre-empt or neutralise the strategy of competitors, or to advance monopoly power, as a defensive strategy to reduce competition.
- To better secure contracts from foreign governments who favour local firms, to better deal with local suppliers and/or labour unions.
- As an initial entry strategy in unfamiliar markets.
- To reduce cross-border political risks.

In most industries we can therefore observe a high degree of concentration. When we know that it has been estimated that the 500 large triad-based TNCs account for 80 per cent of the world's stock of FDI and of over half of world trade (Rugman, 1999: 63), it is obvious that such oligopolistic markets call for appropriate regulation at the international level. Yet, there is no such authority today.

Moreover, in the public sector, we observe exactly the same trend: thanks to liberalisation, deregulation and privatisation, especially in the infrastructure sectors, one can observe the rise of TNCs in sectors such as energy, telecommunications, logistics (transport), postal services, audiovisuals and water, as illustrated in Table 4.4. One can indeed notice that these types of firms are gaining more and more power and importance at the international level.

Telecommunications operators appear to be the most important public services TNCs among the Fortune 500 list of 2000, as shown in Table 4.5.

Table 4.1 Transactions over $10 billion on the merger/take-over market since 1998

Target	*Country*	*Purchaser*	*Country*	*Sector*	*$ billion*
Mobil Oil	USA	Exxon	USA	Oil	77.0
Citicorp	USA	Travelers Group	USA	Finance	72.6
Ameritech	USA	SBC Communic.	USA	Communic.	62.6
AirTouch	USA	Vodaphone	UK	Communic.	62.0
Bank America	USA	Nations Bank	USA	Finance	61.6
TCI	USA	AT&T	USA	Communic.	53.6
GTE	USA	Bell Atlantic	USA	Communic.	53.4
Amoco	USA	BP	UK	Oil	48.2
Chrysler	USA	Daimler Benz	Germany	Car	40.5
Wells Fargo	USA	Norwest	USA	Finance	34.4
Astra	Sweden	Zeneca	UK	Pharmaceutic.	31.0
Asda	UK	Kingfisher	UK	Distribution	31.0
First Chicago	USA	Banc One	USA	Finance	29.5
Atlantic Richfield	USA	BP	UK	Oil	27.5
Rhône Poulenc	France	Hoechst	Germany	Pharmaceutic.	23.0
Ascend Communic.	USA	Lucent Technol.	USA	Communic.	20.0
Bank Boston	USA	Fleet Financial	USA	Finance	16.0
Bayerische Hyp.	Germany	Bay. Vereinsbank	Germany	Finance	13.0
Banco Central Hispano	Spain	Banco Santander	Spain	Finance	11.0
Petrofina	Belgium	Total	France	Oil	11.0

Source: François Chesnais, *A critical view on the Current Consolidation Fever* <http://attac.org>.

Table 4.2 Newcomers to the world's 100 largest TNCs, 1998

Foreign assets (rank)	*Corporations*	*Industry*
13	Suez Lyonnaise des eaux	Diversified
14	Wal-Mart Stores	Retailing
47	Texas Utilities Company	Utility
48	Coca-Cola Company	Beverages
47	Nortel Netwoks	Telecommunications
52	Telefonica SA	Telecommunications
61	RJR Nabisco Holdings	Food/tobacco
63	Stora Enso Oys	Paper
66	RWE Group	Utility
73	Carrefour SA	Retailing
75	Compart Spa	Food
76	SBC Communications	Telecommunications
81	Southern Company	Utility
89	Broken Hill Proprietary	Steel manufacturing
97	Compaq Computer Corp.	Computers
98	SCA	Paper
99	ALCOA	Aluminium manufacturing

Source: UNCTAD (2000: 77).

Table 4.3 Industry composition of top 100 TNCs, 1990, 1997 and 1998 (number of entries)

Industry	*1990*	*1997*	*1998*
Electronics/electrical equipment/computers	14	18	17
Motor vehicle and parts	13	14	14
Petroleum exploration/refining/distribution and mining	13	13	11
Food/beverages/tobacco	9	8	10
Chemicals	12	8	8
Pharmaceuticals	6	13	8
Diversified	2	7	6
Telecommunications	2	4	6
Trading	7	3	4
Retailing	–	1	3
Utilities	–	–	3
Metals	6	–	2
Media	2	1	2
Construction	4	3	1
Machinery/engineering	3	2	–
Other	7	5	5
Total	**100**	**100**	**100**

Table 4.4 Change in rank of selected public services TNCs between 1999 and 2000 (Fortune 500 list, 2000)

Company	*Year*	*Rank*	*Year*	*Rank*
NTT	1999	18	2000	13
AT&T	1999	30	2000	28
E.ON	1999	46	2000	37
Tokyo Electric Power	1999	52	2000	48
Vivendi	1999	68	2000	53
Enron	1999	85	2000	62
Worldcom	1999	214	2000	79

Table 4.5 Telecommunications TNCs in the Fortune 500 list, 2000

Rank	*Company*	*Global 500 Revenues Rank*	*Rank*	*Company*	*Global 500 Revenues Rank*
1	NTT	13	12	Alcatel	163
2	AT&T	28	13	Telefonica	164
3	SBC Communications	42	14	Sprint	212
4	Deutsche Telekom	77	15	China Telecommunications	236
5	WorldCom	79	16	Cable & Wireless	315
6	Verizon Communications	97	17	DDI	351
7	BT	110	18	US West	371
8	Olivetti	112	19	Vodaphone AirTouch	390
9	France Télécom	118	20	Telstra	433
10	GTE	152	21	Teléfonos De México	482
11	BellSouth	156			

Utilities appear to be the second main group, with seventeen firms among the listed Fortune 500 companies. There are also nine engineering and construction companies in that list, eight railways companies and seven energy companies (Table 4.6). This last figure is of particular interest, as some of these electricity companies are also active in the water sector.

One must therefore conclude that there already is a high concentration in the energy sector, dominated as it is by just three firms. Of course, depending on the categorisation, Vivendi and Bouygues could have also been part of this group. Whether in energy, in engineering or in utilities, we can see the predominance of just three firms, Vivendi, Suez Lyonnaise des Eaux and Bouygues, all ranked among the first 100 Fortune 500 companies. And this trend of concentration among public services TNCs continues, as shown in Table 4.7.

In the following section we will now present the three main water TNCs, namely Suez Lyonnaise des Eaux, Vivendi, and Bouygues/SAUR.

Table 4.6 Rank of energy TNCs in the Fortune 500 list, 2000

Rank	*Company*	*Global 500 Revenues Rank*
1	Enron	62
2	Rwe Group	73
3	Suez Lyonnaise des Eaux	95
4	Dynergy	294
5	TransCanada Pipelines	399
6	OAO Gazprom	403
7	El Paso Energy	465

Table 4.7 Biggest utility mergers and acquisitions in Europe between 1998 and 2000

Acquirer	*Target*	*Value of deal $ billion*	*Date completed*
Veba (Germany)	Viag (Germany)	13.15	16 June 2000
Endesa (Spain)	Iberdrola (Spain)	12.96	Pending
Texas Utilities (USA)	Energy Group (UK)	10.95	19 August 1998
RWE (Germany)	Thames Water (UK)	6.26	29 December 2000
Scottish Hydro-Electric (UK)	Southern Electric (UK)	4.47	7 January 1999
RWE (Germany)	VEW (Germany)	3.41	24 November 2000
WPD Holdings (UK)	Hyder (UK)	3.40	26 October 2000
EDF (France)	London Electricity (UK)	3.20	4 December 1998

Source: *The Economist* (27 January 2001).

Public Services TNCs in the water sector

Europe has both the oldest private water and sewerage companies and the highest penetration of private companies in the water sector in the world. This means that private European water companies has have a unique comparative advantage when opportunities emerge both in Europe and globally. The two biggest such water TNCs, Suez Lyonnaise des Eaux and Vivendi, are French. The long tradition of water management delegated to private companies in France made it possible for them to take an early lead in the privatisation of water and sewerage service contracts in the world. In this section we will take a close look at these two companies and their strategies. Furthermore, we will also examine yet another French water company, SAUR/Bouyges, as well as other old and new private companies operating in the global water sector.

Suez Lyonnaise des Eaux

The group Suez Lyonnaise des Eaux is the result of a merger between two old French entities.[1] The origins of Suez can be traced back the creation by Ferdinand de Lesseps of the Compagnie Universelle du Canal Maritime de Suez, later called the Compagnie de Suez in 1875. Lyonnaise des Eaux was created in 1880 under the initial name of Lyonnaise des Eaux et de l'Eclairage. From its very beginning Lyonnaise des Eaux et de l'Eclairage was involved in the distribution of water, in the production and distribution of gas, including public lighting, and in the production and distribution of electricity. More interestingly, with the development of the French colonial empire, Lyonnaise des Eaux acquired as early as 1914 international experience through its activities in North Africa (Morocco, Tunisia), Central Africa (Togo, Congo), as well as in the Pacific (New Caledonia). By 1958, Lyonnaise des Eaux was already distributing water to more than 300,000 customers. In 1967 the Compagnie Financière de Suez becomes the main shareholder of Lyonnaise des Eaux, which diversified, as of 1974, its activities into water treatment with the acquisition of Degrémont in 1972, as well as waste management with Sita (1979), into heating with a participation in Cofreth (1975), and into funeral services with PGF (1979). In 1986, Lyonnaise des Eaux began to get involved in the communications sector, creating Lyonnaise Communications. While both companies were getting more and more involved at the international level, their complementarity led them to merge in 1997 to become the group Suez Lyonnaise des Eaux under the leadership of Jérôme Monod[2] and Gérard Mestrallet.

1 For the history of the two companies, see <http://www.suez-lyonnaise-eaux.fr/french/group/mileston/mileston.htm>.

2 See: Cans (1994: 77–9) for an interesting biography of Jérôme Monod and the links with the French political parties.

Building on this merger, Suez Lyonnaise des Eaux has substantially diversified, and now has operations in four main sectors, water, energy, waste management and communications. In addition, it has also a large construction division. Having restructured its strategy around these four sectors, 75 per cent of the capital employed in 1999 was already concentrated in these four core businesses.[3] Thus, the group is mainly active in infrastructure services, with the ambitions to become, by 2002, the world leader in private infrastructure services,[4] transforming each of its four main sectors into a global player. With 201,000 employees in 120 countries and revenues of 31.36 billion Euro (1999), the ambitions of the group seem within reach.

Let us now look specifically at each of the four sectors, starting with energy. Here Suez is already strongly represented with Tractabel (of which it has owned since 1999 98.19 per cent of the shares[5]) and Elyo (100 per cent), but the group wishes to further develop electricity and gas systems in Europe, as well as in Latin America and the United States. In 1999, for example, it bought EPON (of which it now owns 80 per cent), which is the first Dutch electricity producer.[6] Already today, Suez Lyonnaise des Eaux is the biggest private electricity producer in Europe, and European deregulation in this sector will almost certainly further enhance Suez's expansion. In short, through Tractabel, Suez now has an excellent position in the electricity market.

Suez Lyonnaise des Eaux also has a large presence in waste services (via Sita and Tractabel), and today is the leader in Europe, Latin America and Asia. This position was made possible as early as 1998 with the acquisition of the international business of US waste operator Browning Ferris Industries (BFI) through Sita.[7] Suez is particularly active in Eastern Europe, as Eastern European countries seek to comply with European directives while at the same time privatising. This was, for example, the case in Slovenia in April 1999 when a Suez Lyonnaise des Eaux named 'Aquasystems' agreed to build a wastewater treatment plant. Eastern Europe is in fact of particular interest to companies like Suez, since many Eastern European countries, like many developing countries, benefit from the help of development agencies, in particular the World Bank. For example, the city of Maribor in Slovenia is to receive an Euro 28.1 million loan by the European Bank for Reconstruction and Development (EBRD), which will be immediately used to pay off the TNC services.[8] But Suez is also expanding in the waste sector in Western Europe, for example when it acquired in 2000 a 58 per cent stake of the

3 Suez Lyonnaise des Eaux, Press Release (16 February 2000).
4 *Suez Lyonnaise des Eaux in 1998*, Suez Lyonnaise des Eaux (1998: 1).
5 Suez Lyonnaise des Eaux, Press Release (7 December 1999).
6 Suez Lyonnaise des Eaux, Press Release (19 November 1999).
7 Suez Lyonnaise des Eaux, Press Release (2 December 1999).
8 'Slovenia: EBRD aid', *Global Water Report*, *Financial Times*, 70 (23 April 1999).

company Renoflex, one of the main waste collectors in Copenhagen. This acquisition reinforces the group's presence in Northern Europe and strong growth is expected owing to the Danish industrial market, as well as anticipated privatisations.[9]

The communications sector in Suez is articulated around three main subsidiaries, which are Lyonnaise Cable (of which Suez owns 76 per cent), M6 (34 per cent) and TPS (25 per cent directly and indirectly 25 per cent via M6). Compared with the other sectors, the Group has a weaker position in the communication sectors, where its activities are mainly limited to France and Belgium. Nevertheless, with Lyonnaise Cable, the Group is France's No.1 cable operator with 700,000 homes connected. Despite its success in France, Lyonnaise Cable seems to have had some difficulties expanding on other markets than the francophone countries. However, one of its assets are TV shopping channels, which will extend their international reach, particularly in Hungary, Argentina and Canada. In the beginning of 2000, the group enlarged its strategy by developing a partnership with retailer Rallye-Casino, in the areas of e-commerce and internet, multimedia programming and TV shopping.[10] The marketing force of Rallye-Casino and the communication tool of the Suez Lyonnaise des Eaux will probably bring in a new dynamics to the communication sector. Moreover, a strategic alliance between Suez Lyonnaise des Eaux and Groupe Arnault has just been established in the area of the internet and communication. The prospect of acquiring an UMTS (Universal Mobile Transmissions Standards) licence in France is one of the main reasons for this partnership.[11]

The last, yet most important sector for us, is water. As stressed above, the French companies have a historical dominance and know-how in this area, which has led them to become the undisputed leaders in the privatised water markets all over the world. According to the head manager of this sector, Gerard Payen, only 5 per cent of the water resources are currently managed by the private sector,[12] thus implying that the potential in the water business is enormous. Nevertheless, the group's global presence is already quite impressive. Suez Lyonnaise des Eaux is indeed the world leader in water management businesses, with 77 million people supplied with drinking-water, 52 million supplied with wastewater services and revenues of over 5.12 billion Euro in 1998. If water distribution is ensured through Lyonnaise des Eaux, water treatment engineering is done by its subsidiary Degrémont. Since 1999,

9 Suez Lyonnaise des Eaux, Press Release (9 February 2000).

10 See Suez Lyonnaise des Eaux, Press Release (16 March 2000).

11 'Groupe Arnault et Suez Lyonnaise des Eaux forment une alliance stratégique dans les métiers de l'Internet et de la communication', Suez Lyonnaise des Eaux, Press Release (20 November 2000).

12 'France/Companies: Focusing on water', *Global Water Report, Financial Times*, 90 (2 March 2000).

the group is broadening its strategy, thus regrouping all the different water sector activities into one single unit. Consequently, the Degrémont group brought together in 1999 four specialised subsidiaries (Degrémont Erpac and Joud in France, Schilling Chemie in Germany and the UK's Houseman) to form Aquazur, thus giving rise to Europe's largest specialised water chemicals and environmental services, and the fifth biggest water treatment company in the world. Following the creation of Aquazur, the chairman of this new company, Pierre Thibaudat, declared: 'This merger means that for the first time a European Company will be able to take on the might of US chemical giants Nalco and Betz Dearborn.'[13] Although primarily focused on the European market, Aquazur operates in more than 65 countries, providing a wide range of services from the design and construction of desalination plants to the development and use of water treatment chemicals, but including even power generation, paper production and wastewater treatment.[14] Several months later, the Group Suez Lyonnaise des Eaux launched a take-over bid of US$4.1 billion over Nalco, the No. 1 group in water packaging. Holding 97.1 per cent of Nalco's equity,[15] the Suez group has now brought together Nalco, Calgon and Aquazur, respectively world No. 1, 3 and 5 in water packaging, into a new group based in Chicago under the responsibility of Christian Maurin, president of Degrémont. This strategy brings the group the undisputed leadership in water packaging, with 9600 employees, a turnover of US$2.4 billion and a portfolio of 60,000 industrial clients all over the world.[16]

These moves actually constitute the first step towards a new water strategy. In February–March 2000, the group Suez Lyonnaise des Eaux announced a new structure in its water activities, regrouping all of them into one core business, aimed at creating world leadership in water services, with an expected 60 per cent increase in its turnover by 2004.[17] To this end, the Group has consolidated its water businesses, i.e. Lyonnaise des Eaux, Nalco and Degrémont, into a single unit. In explaining this move, in an interview with the Global Water Report, Philippe Brongniart, a member of Suez Lyonnaise des Eaux's managing committee, said that 'the various strands of the water business were moving closer because industry is outsourcing its water supply and wastewater treatment, and because concerns are growing about fresh water supplies. Equally, there is more interest in treating rainwater, and discussion regarding sludge treatment.'[18] This new concentration of activities

13 'France/Companies: Aquazur takes on US giants', *Global Water Report*, *Financial Times*, 71 (7 May 1999).
14 'France/Companies: Aquazur takes on US giants', ibid.
15 Suez Lyonnaise des Eaux, Press Release (9 November 1999).
16 Suez Lyonnaise des Eaux, Press Release (25 October 1999).
17 Suez Lyonnaise des Eaux, Press Release (24 February 2000).
18 'France/Companies: Focusing on water', *Global Water Report*, *Financial Times*, 90 (2 March 2000).

will bring the group to supply 100 million people and 60,000 industrial clients. It realises a turnover of 8.1 billion Euro, with a growth rate of more than 23 per cent as compared to 1998. This core business activity has under its control more than 60,000 employees in over 120 countries. In this strategy, Lyonnaise des Eaux will be in charge of water resources management, water treatment and wastewater and sludge treatment, while Degrémont will ensure the conception and the construction of water treatment plants and water packaging. But the group has also regrouped its activities in the water sector in order to develop new markets. Thierry Bourbiés' nomination as an executive responsible for the development of water and the internet is one part of this strategy, but the group also wants to invest in emerging markets, such as industrial outsourcing. To this effect, the group has created a Water Resources Advisory Committee (WRAC) to investigate new methods linking treatment, transport and storage infrastructures with the control of consumed volumes, as for example in irrigation and in the treatment of wastewater. Environment also constitutes a growing concern, especially in the field of new technical innovations, such as the collecting and treatment of pluvial waters, but also in the field of managing sludge treatment.[19] Moreover, with the merger of La Lyonnaise Château d'Eau and Roche Claire, the Group has also anticipated the important growth of the supply of refrigerated drinking water fountains, a market which is growing at a rate of some 30 per cent a year in France. Finally, the group has decided to develop e-commerce in all its activities, bringing gains in productivity, rapidity, and efficiency.

The activities of Lyonnaise des Eaux in 1999 have been flourishing, although we should be reminded that a few problems occurred while co-operating with different governments and especially local representatives. In its home market, a few cases of corruption have tarnished Suez Lyonnaise des Eaux's image. This was the case, for example, in Grenoble where the dispute between the city of Grenoble and Suez Lyonnaise des Eaux led the mayor, Michel Destod, to renegotiate the contract. The city bought 16 per cent of Société des Eaux de Grenoble (SEG) from Suez Lyonnaise des Eaux, thus raising its own holding to 67.05 per cent, enough to override the company's objections in any dispute.[20] Having 45 per cent of its water business in France,[21] the Group has to be careful in this market, since it is probably from the revenues in the French market that Suez has managed to develop so quickly internationally. Indeed, the French TNCs benefit from a very protected market where they are the only competitors nearly bringing monopolistic control over the market and fixing prices almost at their

19 Suez Lyonnaise des Eaux, Press Release (24 February 2000).

20 'France/Legislation: Grenoble takes control', *Global Water Report*, *Financial Times*, 69 (1 April 1999).

21 Suez-Lyonnaise des Eaux, Annual Report 1998: 68, 'Geographical breakdown of activities'.

discretion. Moreover, Suez Lyonnaise des Eaux's acquisition of the Générale de Belgique in May 1998 concentrates over 45 per cent of the assets in Belgium, thus creating a binational group seen by most specialists as a dangerous novelty.[22]

At the international level, the group has learned from its past experiences with the controversy over the Casablanca case, where the initial plan to transfer the management of Casablanca water utilities to Lyonnaise des Eaux was highly disputed at the domestic level and also at the international level. It was in 1996 that Lyonnaise des Eaux was awarded a contract to manage Casablanca's water supply, sewerage, and power networks. The controversy was not so much about the fact that the water industry was privatised, as the previous Régie Autonome de Distribution (RAD) was inefficient in financial terms, but it was more about the way the contract was awarded. A diplomat said that 'the Lyonnaise contract was originally conceived after the direct intervention of King Hassan'.[23] As further stressed in the *Middle East Business Weekly*, 'without a competitive bidding process, critics are sceptical that the government has managed to obtain the best possible deal for the Casablancan consumer. The press, businessmen all accuse Lyonnaise of planning to pass much of the investment burden on to the consumer, and of proposing unfair profits'.[24] Since this affair, Lyonnaise des Eaux has been more careful and more transparent throughout the usual bidding process. In Great Britain, the British regulator, OFWAT, announced at the end of 1999 that it had decided to lower the water prices for the next five years (2000–2005). This shows that where there is a regulator, TNCs like Suez have a tougher life. This decision had indeed some serious consequence for Lyonnaise des Eaux, as it holds 70 per cent of the capital of the Northumbrian Water Group. Nevertheless, even with a strong regulator, Suez was able to compromise and obtain a reduction of 13.8 per cent instead of 19 per cent for Essex and Suffolk Water and 19.4 per cent instead of 25 per cent for Northumbrian Water Limited.[25]

Besides these problems, Suez Lyonnaise des Eaux has managed to acquire important contracts in the Middle East and North Africa. For example, in 1999 Dégremont, Suez Lyonnaise des Eaux's subsidiary, won the largest contract it had obtained so far internationally, worth some 1 billion French Francs (approximately 152 million Euro). Indeed, in Egypt, the group builds the second phase of the Gabal el Asfar wastewater treatment plant serving the Egyptian capital, Cairo. In the same year Suez also won an important contract

22 'Suez-Lyonnaise des eaux achève sa mutation en absorbant la Générale de Belgique', *Le Monde* (Tuesday 19 May 1999: 19).

23 'No smooth course for private water', *Middle East Business Weekly*, 41 (10) (7 March 1997: 2).

24 'No smooth course for private water', *Middle East Business Weekly*, ibid.

25 Suez Lyonnaise des Eaux, Press Release (25 November 1999).

for Jordan's capital city, Amman.[26] Two months after winning this contract, the World Bank informed that it had approved a $55 million loan to help fund a $136 million project in order to improve the efficiency, management, operation and delivery of water and wastewater services. As reported by the Global Water Report, 'the bank specifically said that the loan was linked to the recent award of a 51-month management contract secured through competitive bidding by France's Lyonnaise des Eaux, with the UK's Montgomery Watson and the local Jardaneh Arabtech as minority partners'.[27] This constitutes another example of the World Bank and TNCs – in this case Suez – partnership in developing countries. Another interesting point is that this contract reflects the new strategy of the group, where the distribution and wastewater services are increasingly combined.

Another successful strategy is the use of its subsidiary, Aquas de Barcelona (Agbar), in order to gain rapid market access in Latin America. In June 1999, for example, the group was successfully awarded a major contract in Santiago, Chile. This concession is designed for an unlimited period and concerns 5 million inhabitants regrouped into 44 districts.[28] And this contract is yet another illustration of the group's new strategy combining water distribution and water treatment. The group is also present in Buenos Aires, Cordoba and Santa Fe in Argentina, in La Paz and El Alto in Bolivia, in Limeira in Brazil and in Bogota in Colombia. This region of the world therefore represents an important region of the group's world strategy, currently serving 16 million inhabitants in the distribution of water and 11 million in water treatment.[29]

But ultimately, the Group's real interest may be Europe, where the biggest markets lie. Recent developments in Europe are thus of particular interest, and this is especially the case for Germany, which is experiencing the first wave of privatisation. Until now, the German market had been very protected, one of the reasons being that their model of water resources management has been considered until very recently as being one of the best.[30] However, Suez Lyonnaise des Eaux has actually managed to enter the German market through its subsidiary Eurawasser. In 1990, the group was present in Rostock and Postdam, where it distributed and treated the water. It is also responsible for the water treatment of the town of Goslar. In 1999, the group was awarded another contract in the Saxe-Anhalt Land for the distribution of water for more than 50,000 people. The contract is for a 25-year period and represents

26 For more details, see Suez Lyonnaise des Eaux, Press Release (2 February 1999), 'Jordan/ Infrastructures: French win in Amman', *Global Water Report*, *Financial Times*, 65 (5 February 1999).

27 'Jordan/Funding: Amman gains $55m World bank loan', *Global Water Report*, *Financial Times*, 69 (1 April 1999).

28 Suez Lyonnaise des Eaux, Press Release (14 June 1999).

29 Suez Lyonnaise des Eaux, Press Release (14 June 1999).

30 See Briscoe (1995b: 15).

a turnover of 34 million French Francs per year (approximately 5.19 million Euro).[31] However, it should be emphasised that Eurawasser operates through a consortium, of which it holds only 25 per cent of the shares, while the municipalities have a majority of 75 per cent.[32] The group's expansion in Europe can also be witnessed in the case of the second water treatment contract awarded to Lyonnaise des Eaux and Dégremont in Scotland under a build–operate–transfer (BOT) agreement for the region of Ayrshire. The contract represents an investment of more than 497 million French Francs (approximately 75.8 million Euro) and will concern the treatment of water for 450,000 people. The selected consortium is a partnership between the Northumbrian Water Group (Lyonnaise des Eaux's British subsidiary), Dégremont and AMEC, which will ensure the financing, the conceptualisation, the construction, as well as the management of three treatment plants and a sludge treatment centre over a period of thirty years.[33]

Finally, in 1999 the group has expanded into the central European market and more especially the Czech Republic. It has renewed its contract in Brno and has been awarded a contract in Horni Slovak. In fact, Lyonnaise des Eaux has been present in the Czech Republic since 1992. The group's presence in the distribution of water, as well as water treatment, represents a turnover of 117 million Euro and ensures services for a population of 2.1 million located in Brno, in Karlovi-Vary, in the south of Moravia and in the Ostrava region.[34] Another success for the Group in this region is its contract with the municipality of Trencin in Slovakia. According to a press release,[35] this contract is a significant step for the group to develop into Central Europe where it already serves 4 million customers in the distribution of water and 2 million in the treatment of water.

Vivendi

Many similarities can be found between Vivendi and Suez Lyonnaise des Eaux. Indeed, both have a long tradition in public services provision. However, compared to Suez Lyonnaise des Eaux, Vivendi did not diversify its activities until 1967. Vivendi was founded in 1853 under the name of Compagnie Générale des Eaux. It started supplying water to Lyon in 1853, and expanded its activities to Nantes in 1854 and to Paris in 1860. In the 1880s, the firm had already acquired an international experience when international treaties awarded the production of water to Générale des Eaux in 1880, followed by two major contracts in Constantinople in 1882 and in Oporto in

31 Suez Lyonnaise des Eaux, Press Release (21 June 1999).
32 Suez Lyonnaise des Eaux, Press Release (21 June 1999).
33 Suez Lyonnaise des Eaux, Press Release (2 September 1999).
34 Suez Lyonnaise des Eaux, Press Release (7 October 1999).
35 Suez Lyonnaise des Eaux, Press Release (20 October 1999).

1883. However, unlike the other two French public service TNCs, Générale des Eaux has, until the early 1960s, limited its activities at the national level and to the water sector. It was only in 1967 that it decided to change its strategy and move into the management of district heating and the operation of household waste incineration and composting plants. In 1976, water distribution still accounted for the bulk of Vivendi's net sales (46 per cent), while civil engineering and construction accounted for 28 per cent and energy for 12 per cent. It is from the 1980s onwards that we witness a rapid growth when Générale des Eaux became France's leading private energy management company by increasing its holdings in the Compagnie Générale de Chauffe to 100 per cent. The same year, the group broadened its services into wastewater treatment with the creation of Omnium de Traitement et de Valorisation (OTV), and into waste management and transport by taking over Compagnie Générale d'Entreprises Automobiles (CGEA). In the 1980s, the group started to enter the audiovisuals and the telecommunications sector with the Havas media group by creating the pay-TV channel Canal+ in 1983 and SFR in 1987. The next year, under an agreement with Saint Gobain, Générale des Eaux entered the construction sector, taking shares in the construction and civil engineering giant, Société Générale d'Entreprises (SGE). Gradually, the group became what we can call a 'public service TNC', encompassing all infrastructure domains from water distribution to waste management and from electricity to telecommunications. In addition to diversification, the group also expanded internationally. This was especially the case during the early 1990s where Générale des Eaux started to operate water services in the United States, in the Asia Pacific region (Sydney, Kuala Lumpur and Tianjin, China's third biggest city) and in Latin America (Mexico City and Caracas). Its European water business is now particularly strong in the United Kingdom and Spain. The end of the 1990s witnessed yet another step in Vivendi's expansion, as the name Générale des Eaux was transferred to the water business, and the group became Vivendi.[36]

Vivendi is today one of the largest public service TNCs,[37] with 235,610 employees in 1998. The results of the first half of 2000 showed strong growth, with a 9 per cent increase in its net sales as compared to the first half of 1999.[38] At the end of the 1990s, Vivendi seemed to have refocused most of its activities around two core sectors, environment and communication. While the group had consolidated net sales of 31.7 billion Euros, these two sectors

36 Most of the facts quoted in this paragraph come from the Company's history webpage. For more details, see <http://www.vivendi.com/en/html/groupe/repere1.cfm> (and <. . ./repere2.cfm> and <. . ./repere3.cfm>).

37 See <http://www.vivendi.com/en/html/pratique/webs.cfm>. This site shows all the major firms under Vivendi's control.

38 'Vivendi: net sales for the first half of 2000', Vivendi Press Release (4 August 2000).

represented respectively 15.4 billion Euro[39] and 5.96 billion Euro, accounting for more than 67.4 per cent of the group's activities. However, it should be recalled that up to 1998 slightly less than 50 per cent of Vivendi's activities were in the environmental sector and that the communications sector was the least developed. The three main activities then were utilities, construction and property, and communication. While it is clear that Vivendi is a TNC, one must also note that the net sales outside of France make up about only a third of its net sales. This shows, that for Vivendi, as well as the other French TNCs, France remains the core market, where they can benefit from oligopolistic power. Overall, 70 per cent of Vivendi's revenues are in Europe. After France, the United Kingdom constitutes the best market, followed by Germany and Spain. The group's penetration of the Eastern market is in fact not yet significant. Surprisingly, North and South America are the second biggest markets totalling 1466 million Euros (as compared to 7741 million Euros for Europe). Africa and the Middle East constitute the third biggest market, with Africa being predominant. This can be explained by the serious water scarcity in the Middle East and the very few interesting markets as yet open for competition there. The last region, Asia Pacific, does not yet generate important revenues so far despite an aggressive policy developed in many countries, in particular Australia and China, but also Cambodia, Hong Kong, Indonesia, Malaysia, New Caledonia, New Zealand, Pakistan, the Philippines, Singapore, South Korea, Taiwan, Thailand and Vietnam.

'Stable economies'

Vivendi defines its strategy around the concept of 'stable economies'. According to the Vivendi's annual report, 'stable economies account for 95 per cent of business'.[40] When presenting its strategic priorities for 1998 and 1999, Vivendi strongly stresses this idea of stability. Indeed, the group wishes to strengthen its operations in the United States and has been, up to now, quite successful in penetrating the US market. It also follows Suez Lyonnaise des Eaux in its attempts to penetrate the Spanish-speaking markets, including Spain. As we will see later, the group has established a solid base in Spanish utilities with FCC and in Spanish communications with Doyma and Anaya. This argument for stability is probably a response to past criticism of the group's rapid diversification. Indeed, until recently, Vivendi comprised more than 25 different activities.

39 In Vivendi's 1998 *Annual Report*, this sector is still called 'utilities' rather than 'environment'. This can be explained by the fact that the strategy changed only in 1999.

40 Vivendi, 'The year in brief, 1998, Utilities, Construction and Property, Communications, Vivendi', *Annual Report, 1998*: 6.

Communications

As noted above, the group now wishes to concentrate on two sectors, environment and communications. Since the beginning of 2000, there are strong indications that the group wishes to reinforce its communications sector. In March 2000, CEO Messier indicated that there would be a separate listing of its utilities and of its communications businesses on the New York Stock Exchange. Mr Messier reported that the move would 'clarify things for the shareholder whose investment priorities is communications', further explaining that this marked 'the final stage of Vivendi's restructuring'.[41] Although Vivendi's spokesman indicated that the revenues from the environmental business would not be used for the development of telecommunications,[42] it was reported that Vivendi's environment division would assume 16.5 billion Euros of the group's debt, leaving the communications division (as well as the holding company) debt-free.[43] This move clearly indicates the group's willingness to focus more specifically on communications, while receiving the easy benefits of its environmental services resulting from the quasi-monopolistic character of these markets. One can actually note that this approach is not favoured by financial analysts, as Vivendi Environment is rated only BBB+ for the long term and A-2 for the short term, as compared to Suez Lyonnaise des Eaux whose ratings are, respectively, A and A-1.[44] In the media and communications sector, Vivendi has under its control the French publisher Havas, pay-TV Canal+, and telephone operator Cegetel. In the beginning of 2000, Vivendi moved into the mobile phone sector through a joint venture with Vodaphone. For the group, this acquisition reflected the importance given to communication as the business with the greatest growth potential. The collaboration with Vodafone was done skilfully by playing Vodafone against Mannesmann.[45] However, most analysts remain wary regarding Vivendi's chances in the telecommunications sector. Nevertheless, Vivendi's ambitions were reinforced with the creation of Vivendi Universal, resulting from the strategic business combination between Vivendi, Canal+ and the Seagram Company Ltd. As Vivendi stated,

> Vivendi Universal will be ideally positioned to accelerate the world-wide growth of wired and wireless Internet access devices by providing proprietary, value-added content, e-services and e-commerce to customers

41 'France/Companies: Vivendi separates listings', *Global Water Report, Financial Times*, 90 (2 March 2000).

42 'France/Companies: Vivendi separates listings', ibid.

43 'Environment supports debt', *Global Water Report, Financial Times*, 89 (14 February 2000).

44 *Notations, La référence sur le risque de crédit*, 50 (September 2000), Standard & Poor's SA.

45 For more details, see 'Praise for Messier', *Global Water Report, Financial Times*, 89 (14 February 2000).

anytime, anywhere. The combined company will bring content from the world's largest music company, second largest film library, major film production studio, second largest destination theme park company, and global leader in reference, consumer and PC-based software game publishing with Vizzavi, Vivendi's new multiple access portal, and the combined global distribution capabilities of Vivendi, Seagram and Canal+.[46]

Environment

In the environmental sector, the group is one of the world's leaders. Vivendi provides water and wastewater services through Générale des Eaux, produces electricity through Sithe Energies,[47] and manages heating networks through Dhakia. It collects, recovers and recycles municipal and industrial waste through Onyx, and operates transport network through CGEA Transport. With this network, the group presents itself as the world leader in utilities. Transport is the smallest sector, representing 1.99 billion Euro, with most of the sales outside France. This activity has basically developed in two main regions. In Europe, the group has considerably expanded in operating rail services in the United Kingdom, the Netherlands, Portugal and Sweden. Combining rail and bus transport, CGEA Transport carries 600 million passengers a year in Europe, making it Europe's largest private operator. The most important market is in the United Kingdom where the group, together with Connex Rail, manages almost 20 per cent of the British rail network through two franchises (South Central and South Eastern), both won in 1996.[48] The other main successes of the group are in the Asian Pacific region and especially in Australia, where it operates a monorail and a light rail express in Sydney and a bus company in Perth.[49] Several projects are also being examined for Manila in the Philippines, Malaysia and Indonesia.

Energy

As for the energy sector, the group declared in 1998 its ambition to make this sector one of its future major business activities. It is in this perspective that on 5 October 1998 Dalkia was launched, the new name for Vivendi's energy

46 'Vivendi, Seagram and Canal+ to merge, creating a fully integrated global media and communications company for the wired and wireless world' (20 June 2000), Vivendi Press Releases <http://www.vivendi.fr/fr/html/communiques/200600_lang.html>.

47 One should note that Vivendi has now only a minority holding of about 30 per cent in Sithe Energies Inc., after the Group finalised a contract with PECO Energy Company (Vivendi Press Release, 14 August 2000).

48 For more details, see 'Vivendi, Business activities in the UK', *Thema*, 1 (June 1998: 38–43).

49 For more details, see 'Vivendi, Vivendi's activities in Asia Pacific', *Thema*, 2 (December 1998: 53–6).

services. This move brought all the different subsidiaries pertaining to energy into one single structure. Eastern Europe constitutes a considerable potential market for Dalkia, which has already obtained contracts in Hungary, the Czech Republic and Poland. Vivendi has also attacked the US market with Sithe Energies,[50] especially in the Northeast where it controls a market of over 70 million inhabitants. Reflecting the same level of activity as the energy sector,[51] Vivendi's waste management companies are successfully expanding their activities, since their alliance with FCC of Spain. As underlined in the 1998 *Annual Report*, this strategy has enabled the group to win contracts in Australia and in North and South America. Acting as partners of local authorities, the companies CGEA-Onyx, Sarp and Sarp Industries regrouped 42,000 employees and managed the waste of 30 million people worldwide. Moreover, 53 per cent of the net sales in the energy sector are generated in industry.[52]

Water

As part of the environment division, the water sector constitutes the largest share, representing net sales of 6.73 billion Euro, the equivalent of the waste management and the energy sectors together. As underlined above, water was the initial sector of the group and therefore benefits from a historical domination and know-how, making Vivendi, together with Suez Lyonnaise des Eaux, the undisputed world leaders. This fact was revealed when Vivendi purchased USFilter and started to develop an aggressive policy on the US market. With this acquisition, it was estimated that the environment businesses would have a turnover of some 22 billion Euro ($US23.7 billion).[53] This represented the largest French acquisition ever in the United States, and was estimated to double the revenues of Vivendi's water treatment business. Combined, USFilter and Générale des Eaux would have an annual sale of approximately US$12 billion (11.0 billion Euro).[54] Moreover, according to both companies, cost-saving innovations in seawater desalination technology would double the size of the worldwide desalination market to more than $70 billion (66 billion Euro) during the next twenty years. Approximately $10 billion (9.4 billion Euro) in new desalination projects are scheduled to

50 One should note, as previously mentioned that it is now the PECO Energy Company which owns 49.9 per cent of Sithe North America (Vivendi Press Release, 14 August 2000).

51 Net sales over FF20 billion (FF23.4 for the energy sector and FF20.4 for waste management).

52 Vivendi, Annual Report, ibid: 30.

53 'France/Companies: Vivendi finances USFilter purchase', *Global Water Report, Financial Times*, 70 (23 April 1999).

54 'US/Merger & Acquisitions: Vivendi grabs US headlines', *Global Water Report, Financial Times*, 69 (1 April 1999).

come on line worldwide in 2000–2005.[55] However, the most immediate result is that this important acquisition gives Vivendi an important place on the US market. USFilter has been a successful company, boosting its profits from $417 million in 1993 to around $4.4 billion (approximately 4.1 billion Euro) in 1998.[56] Since 1991, it has bought more than 150 companies, thus regrouping complementary skills including equipment, service provision, and engineering solutions. Besides the American market, Vivendi obtained many successes in the water sector during 1999–2000. In Latin America for example, the group, through CGE-Sahide, won a substantial increase in the size and the scope of their ongoing contract with the Puerto Rico Aqueduct & Sewer Authority.[57] On the Asian market, China constitutes a new important market. Vivendi, with Marubeni, sponsored the Chengdu BOT scheme, representing $106.5 million. In Turkey, OTV won a $11 million contract to construct a sewerage collection and treatment system in Midas in the south west of the country. In the United Kingdom, through General Utilities, the group managed to acquire a considerable part of the market. In April 1999, South Staffordshire Water[58] managed to expand its activities with the £3.2 million purchase of Epok, a drain and sewer cleaning company. In Switzerland, the group (essentially through Onyx Suisse) is gradually penetrating the market. Besides the Group's activities in waste management, Vivendi Water is helping in the renovation of water treatment plant and obtained in January 2000 a 70 million Swiss Franc contract to renovate Bern's water treatment plant.[59]

In short, Vivendi, like Suez Lyonnaise, is strongly expanding not only across sectors but also globally. Both are active in approximately the same sectors, i.e. public services at large, even though Vivendi seems to have a clearer strategy and a stronger presence in the area of telecommunications and the new internet economy. Both also substantially use subsidiaries in order to expand in the different continents and countries.

Bouygues/SAUR

Like French Vivendi and Suez Lyonnaise des Eaux, Bouygues is one of the largest public services TNC in the world. Bouygues is well known in the construction sector, and is today recognised by most specialists as the biggest construction company in Europe, if not the world. However, the Bouygues Group is in fact as diversified as are Vivendi and Suez. In 1998 the Bouygues Group recorded a turnover of 15.2 billion Euro, covering eighty countries.

55 See <http://www.usfilter.com> (25 August 1999).
56 'US/Merger & Acquisitions: Vivendi grabs US headlines', ibid.
57 'Puerto Rico–PRASA deal', *Global Water Report*, *Financial Times*, 67 (5 March 1999).
58 27.6 per cent of its capital belongs to General Utilities ('UK: South Staffs purchase', *Global Water Report*, *Financial Times*, 70 (23 April 1999).
59 'Vivendi vient poutzer la Riviera', *L'Hebdo*, 17 February 2000: 43.

The international character of this firm is well established, with around 50 per cent of its workforce abroad. One characteristic of the Bouygues group, as compared to the other two public services TNCs, is the regional structure of its workforce. A little more than 50,000 employees are located in France, while the second largest regional presence is in Africa, rather than in Western Europe, with 22,525 workers as compared to 10,724 in Western Europe. This involvement on the African continent has led many authors to consider Bouygues as colonialists, which was brilliantly shown in Smith and Glaser's book, *Ces Messieurs Afrique*.[60] Like the two other French TNCs, the group went through a substantial restructuring process when, in 1998, it decided to focus on two main sectors, thus asserting its identity as a construction and as services group.

Construction

The construction sector is Bouygues' original core activity, and as such is divided into three sub-sectors, building and civil works, property, and roads. The group holds leading positions in these three sectors, which recorded a total turnover of 10.4 billion Euro in 1998. In comparison to the other two sub-sectors, it is the construction sector that is the most important activity of this group. It represents 69 per cent of total turnover with three different firms, BTP for building and civil works, Bouygues immobilier for property, and Colas for roadworks. Colas is the world leader in roadworks and is present in forty countries. In terms of property, the market is essentially French and to a certain extent European, with some presence in Spain and Germany. The building and civil works sector has a strong international character, with 51 per cent of its turnover abroad. This sector had a turnover of 5.2 billion Euro and executed the major contracts which gave Bouygues its high reputation.

Communications and utilities

Although the group has identified two sectors, one can see that there are in fact three different core business activities. Besides construction, we can indeed find, under the heading of 'services', two different sectors, i.e. communications and utilities. In the communications sector, the group is in direct competition with the other two French TNCs, owning the media channel TF1. However, we can see that, compared to M6 (Suez) and Canal+ (Vivendi), TF1 is the leader with 35.3 per cent of the audience ratings, as compared to 12.9 per cent for M6 and 4.6 per cent for Canal+. TF1 has also some thematic channels under its control, such as Eurosport and LC1. Its publishing and distribution branch, with TV shopping and TV videos, constitutes an

60 Smith and Glaser (1992).

important income for TF1. In terms of telecommunications, the group, with Bouygues Telecom, also is confronted by another French TNC, Vivendi's SFR. Bouygues Telecom has been very successful since it was launched in 1996. According to its own sources, Bouygues Telecom has performed better than any other GSM 1800 operator worldwide thirty months after launching, in terms of numbers of clients. The French cellular phone market is the biggest market in Europe, with Itinéris and SFR having over 2,500,000 users fifty months after launching and Bouygues Telecom having over 1,700,000 users thirty months after launching. Telecommunication represents 15 per cent of the Group's activities.

In utilities management, the group is represented by Société d'Aménagement Urbain et Rurale (SAUR). With over 22,200 employees, SAUR's turnover represents 14 per cent of the group's total. The SAUR group is involved in the water sector, in refuse collection, in waste management, as well as in energy distribution and generation. In 1998, having failed to find a partner for SAUR, Bouygues decided to restructure SAUR's activities. Since December 1998, SAUR now owns 100 per cent of SAUR International while Electricité de France International swapped its 22.7 per cent holdings in SAUR International for a 14 per cent holding in SAUR.[61] The net result is that SAUR has a quite simple ownership structure with Bouygues holding 73 per cent, EdF International 14 per cent and an investment fund 13 per cent. The international activity of SAUR is a high proportion of its total business. Indeed, Saur's turnover in 1998 was 2.1 billion Euro, of which 33 per cent was from abroad. In France, usually seen as the third major actor, SAUR serves 2.9 million customers. In its international activities, SAUR has been criticised for its colonial attitudes, as more than half of all its international business is in Africa (66.7 per cent in 1998), and more specifically in Francophone Africa. However, many water specialists see SAUR International's subsidiary, SODECI, as the most successful model of water resources management in Africa. In terms of sectoral activity, SAUR has, like the two other French TNCs, diversified its activities, in water, in cleansing, in energy and in multi-service contracts. Multi-services contracts are a relatively new market, providing management services for community facilities (e.g. catering, transport, parks), as well as technical management of public buildings, school premises and hospitals, for example.

Water

There is also another characteristic of the SAUR group, as compared to the two other French TNCs. Unlike the other two, the water sector for SAUR

61 'France/Companies: SAUR restructures', *Global Water Report*, *Financial Times*, 63 (9 January 1999).

represents 77 per cent of its 2.1 billion Euro turnover. SAUR can thus be seen as one of the emerging 'rivals' of both Vivendi and Suez Lyonnaise des Eaux in this sector. Interestingly, as is the case with the two other French TNCs, the Spanish and Latin American markets seem also of strategic interest. SAUR has three subsidiaries in Spain, Gestagua, Aquas de Valencia and Emalsa, from which an expansion into Latin America in the near future would not come as a surprise. In the United Kingdom, the Group has also a solid market with SAUR UK, especially in the South of England. Another European market of particular interest for SAUR is Italy. SAUR has been present in Italy since 1997 when it acquired Sigesa from the Fiat Group. At the beginning of 2000, SAUR challenged the other two French TNCs by merging with the Italian construction group Italomobiliare, providing water distribution and treatment operations in Italy through Italomobiliare's Crea subsidiary. Sigesa will take a 71 per cent holding in Crea while Italomobiliare will retain 29 per cent. Crea currently provides water in thirteen regions, and gas in three regions of Italy. It supplies drinking water for some 840,000 people, and processes the wastewater generated by the equivalent of 2 million inhabitants.[62] Together with Sigesa, the new group will provide drinking water for 1.2 million people, and provide sewerage treatment for 2.8 million inhabitants. A turnover of some 100 million Euro will make Crea the second largest private water services provider in Italy.[63] In commenting on this contract, a SAUR spokesman told the *Global Water Report* that the Italian market was particularly active, given the creations of water basins. He added that 'the collaboration with Crea was typical of the way SAUR developed internationally. Effectively, it was a partnership with a national group, making the new group Italian as well. This would allow the group to offer better services because it knew the local market and local customs', concluding with the fact that 'local content was a constant factor of SAUR's policy'.[64] The Group was also active in 1999–2000 in Africa, where it was awarded a contract in South Africa,[65] and is in the bidding process with Vivendi in Tanzania.[66]

Although we can see that in the domain of utilities, Suez Lyonnaise des Eaux and Vivendi have developed a far more aggressive strategy than the Bouygues Group with their diversification and internationalisation, these three TNCs have many similarities. Their success lies in their historic

62 'Italy/Companies: SAUR enters Italy', *Global Water Report, Financial Times*, 89 (14 February 2000).

63 'Italy/Companies: SAUR enters Italy', ibid.

64 'Lively market', *Global Water Report, Financial Times*, 89 (14 February 2000).

65 See 'South Africa: SAUR wins Dolphin Coast', *Global Water Report, Financial Times*, 65 (5 February 1999).

66 See 'Tanzania/Privatisation: French frontrunners for Dar', *Global Water Report, Financial Times*, 89 (14 February 2000).

predominance and technical know-how. However, it seems that their strengths relied, and still rely to a certain extent, on their protected home market. All three groups have been repeatedly fined for cartelisation. For example, the European commission is currently prosecuting the French government over its award of the construction contract for the Stade de France, which it awarded to a Bouygues/Vivendi/Suez Lyonnaise des Eaux joint venture. The French government allegedly ignored EU rules on fair advertising and presented the contract as a concession, not as a simple work contract.[67]

Old and new competitors: not much space left . . .

There are currently not many companies in the area of public utilities which can compete with the three French TNCs just discussed, and this was especially true until very recently in the water sector. The British water industry had some previous experience during the nineteenth century in the water sector. The British companies were the first providers of piped water in London. The English companies were also the first to be granted concessions in other countries, Berlin in 1856 and Cannes in 1866 (Roth, 1987: 251–9). However, the British water industry soon became nationalised (Finger and Lobina, 1999: 21). Nowadays, the situation is very different. The privatisation of the British water industry in the 1980s is too recent for them to compete with the French water TNCs, not to mention the fact that many privatised British companies were directly bought by the three French TNCs mentioned above. This leads some analysts, when they compare UK and French water TNCs (in this case Suez Lyonnaise des Eaux), to say

> Suez makes most of our companies look like corner shops. Not only are they so much larger than the others and on line for a steady profit increase of about 10% going forward, but their thinking seems to be much wider. They see the technology side so clearly – the R & D under their wing is amazing . . . The water division looks extremely positive, especially given the way that privatisation is going to increase. It looks as though they will take more than their market share in this respect.[68]

Nevertheless, the privatisation under the Thatcher government in the early 1980s gave rise to several companies, some of which are now emerging at the international level.

In their development strategy these companies follow, to some respect, the same pattern as the French water TNCs. For example, Yorkshire Water has developed a multi-utility strategy like the French TNCs, declaring that its

67 See <www.psiru.org>, 'Companies information'.

68 'France/Companies: Focusing on water', *Global Water Report*, *Financial Times*, 90 (2 March 2000).

ambitions were as follows: 'We are an infrastructure group operating in Europe and North America. Our activities are expanding and we now have interests in water services, environmental services, waste management, energy and land development.'[69]

At the international level, we can note a recent success in Chile for Anglian Water International when winning a 35 per cent stake in Esval, the water and sewerage company which serves the Valparaiso coastal area of Chile, and this in a bid against Suez Lyonnaise des Eaux and Vivendi.[70] Valpariso's Esval is the second largest water and wastewater supplier in Chile. Esval serves a population of 1.5 million, with 98 per cent of its customers in water, and 90 per cent receiving sewerage services.[71] Anglian Water International is looking for further opportunities in the United States, and aims to win more contracts in China and in New Zealand, for example, where it is already running a twenty-year Design–Build–Operate (DBO) contract for a new wastewater treatment plant serving the capital, Wellington. Another successful British Water TNC is Biwater. However, as we will see later, this group has gone through several crises before having some success. In 1999 Biwater announced a joint venture with Nuon NV, the Dutch multi-utility firm. The deal, which valued the partnership at $260 million (approximately 241 million Euro), will be used to expand Nuon's and Biwater's global range. At present, Biwater supplies water and wastewater services to over 2 million people worldwide. Nuon NV is one of the largest multinational energy and water companies in the Netherlands and has operations in China, the United Kingdom, Israel, the United States and Eastern Europe. Apparently, the joint venture will particularly explore the business opportunities in Eastern Europe.[72] In 1999, Biwater was also awarded a contract in India,[73] a market that is not yet open to competition. Another British company that should be mentioned here is Thames Water. In 1999 it bought the Swansea-based Memtech (UK) Ltd. This latest acquisition adds Memtech to its existing PCI Membranes firm, creating one of the largest membrane systems' companies in Europe. The Managing Director of Thames Water's Products Division, Steve Walker, said that this was part of the company's strategy 'to grow its non-regulated businesses'.[74] Another British water TNC we should mention is United Utilities, operating as North West

69 <http://www.keldagroup.com>.

70 'UK/Companies: Anglia looks ahead', *Global Water Report*, *Financial Times*, 65 (5 February 1999).

71 'Chile/Privatisation: Chilean success for Anglian Water', *Global Water Report*, *Financial Times*, 63 (9 January 1999).

72 'Netherlands/Companies: Biwater to partner Nuon', *Global Water Report*, *Financial Times*, 89 (14 February 2000).

73 See 'India/privatisation: Biwater wins Bangalore BOOT', *Global Water Report*, *Financial Times*, 65 (5 February 1999).

74 'UK: Thames buys Memtech', *Global Water Report*, *Financial Times*, 68 (19 March 1999).

Water (NWW). However, its recent international experience has been slightly disappointing.[75]

Although we can observe a spectacular growth for several British water TNCs on the international market, it should be emphasised that the British firms have also faced many problems. Thames Water's willingness to step into non-regulated business stems from the fact that Britain is considered to be a very highly regulated market, which is precisely why the international market is seen as a much greater opportunity, 'which the UK industry will clearly depend on to counter low returns from their regulated businesses'.[76] This even lead the *Global Water Report* to hint at the difficulties of the British water sector, when saying that 'the days of easy profits for the water companies of England and Wales are becoming a distant memory. Investors believe that however cordial a relationship a utility might have with its regulator, this alone will not deliver shareholder value'.[77]

Investors criticise the lack of expertise in management, relentless downsizing and short-sighted efficiency drives. According to the management consultant AT Kearney, cost reductions are going to become increasingly difficult, with capital expenditure hedged by European legislation, so radical thinking will be necessary to produce longer-term gains.[78] Investors also wish that British firms had a sharper approach to international possibilities. As Kevin Rutledge, one of the report's authors, remarked:

> There are a lot of opportunities around the world, and the water companies need to get better at spotting them and negotiating for contracts. These contracts will require different skills from those needed for running water at home – skills in the field of project finance and project risk evaluation, together with the ability to adopt a range of local cultures. Companies also require the ability to act like a multinational, which some water companies don't have.[79]

The past international experiences of the British water companies show how valid the above analysis is. There are many examples of politicisation over water privatisation and perhaps the inability of British firms, as compared to the French, to deal with these situations. This was the case with

75 For more details, see 'UK/Companies – United Utilities' change of strategy', *Global Water Report*, *Financial Times*, 104 (15 September 2000).

76 'UK/Companies: Anglia looks ahead', *Global Water Report*, *Financial Times*, 65 (5 February 1999).

77 'UK companies: Time for clarity', *Global Water Report*, *Financial Times*, 66 (19 February 1999).

78 'UK companies: Time for clarity', *Global Water Report*, *Financial Times*, 66 (19 February 1999).

79 'UK companies: Time for clarity', *Global Water Report*, *Financial Times*, 66 (19 February 1999).

Thames Water's problems in Malaysia, where a conflict emerged between Kelantan Water, a joint concession with Thames, and the government, who refused to pay for the services.[80] Another British TNC, United Utilities, has been criticised for its performance, pricing and treatment of workers in Manila in the Philippines.[81] The same can be said for Anglian Water, which has been repeatedly unsuccessful in every continent. For example, in 1997, Anglian wrote off £8.2 million by selling loss-making companies. It also wrote off £7.3 million of investments in Brazil, £2.8 million investments in the Czech Republic and £8.7 million in general restructuring charges and provisions on contracts.[82] The most mediated contract was Biwater's South African Nelspruit and Matsulu contract, which has been faced since its inception by very strong opposition from the labour unions.[83]

In terms of domestic performance, there are also many errors in the past management of water resources. Like many other UK water companies, Thames Water has a poor record on leaks and did not cope well with a period of dry weather in England in 1995.[84] As the *London Evening Standard* reported, 'Thames Water suffers more leakages than any other water company in Britain, losing 333 litres for every household in the region.'[85] Thames Water's reputation was also tarnished with several complaints of unfit water. In 1996, the group was fined £80,000 plus costs after being prosecuted for supplying water 'unfit for human consumption'. The *Financial Times* added that it had delivered in 1994 to 84,000 customers water that 'looked very dirty even though there was no health hazards.'[86]

As a result of this poor performance, specialists in the water sector advise companies to follow a new strategy. This strategy consists essentially in seeking out new areas of business, either in non-regulated markets in the United

80 For more details, see 'Thames negotiates over Kelantan problems', *Global Water Report, Financial Times*, 65 (5 February 1999).

81 See PSIRU Summary, United Utilities <http://www.btinternet.com/-ipspr/companyinfo/377/cap.htm>.

82 See PSIRU Summary, Anglian Water, <http://www.btinternet.com/-ipspr/companyinfo/17/cap.htm>.

83 Much literature can be found on the two years of negotiation before Biwater was awarded the contract. See Norm Dixon, '"Odd couple" in South African water privatisation plan', Green Left Weekly Home Page; 'Biwater seeks to suppress public debate and grassroots organizing over the Internet', Corporate Watch Homepage, 'South Africa/Privatisation: Private/public framework signed', *Global Water Report*, *Financial Times*, 63 (9 January 1999); 'South Africa/Privatisation: Nelspruit signing', *Global Water Report*, *Financial Times*, 70 (23 April 1999).

84 For more details, see 'PSIRU Summary, Thames Water', <http://www.btinternet.com/-ipspr/companyinfo/357/cap.htm>.

85 *London Evening Standard* (8 July 1996), quoted in 'PSIRU Summary, Thames Water', <http://www.btinternet.com/-ipspr/companyinfo/357/cap.htm>.

86 *Financial Times* (1 August 1996), quoted in 'PSIRU Summary, Thames Water', <http://www.btinternet.com/-ipspr/companyinfo/357/cap.htm>.

Kingdom or by redoubling their efforts to gain business in overseas markets. The highly fragmented waste sector is seen as having precisely such good prospects for growth. In 1999, for example, Vivendi bought Leigh Environmental, a major UK waste business, and Yorkshire Water performed a complicated manoeuvre by merging its Global Environmental waste interests with Waste Recycling Group for a 46 per cent stake in the company.

Spain

When examining the water sector, one cannot ignore the Spanish water industry. Although we have already examined Aquas de Barcelona and FCC from the French Water TNCs' point of view, there are still several elements that we should add, in particular concerning the role of Aquas de Barcelona (Agbar). Aquas de Barcelona is the largest water company in Spain. It also owns the second largest waste management company, Cespa, and a large private health care company, Adeslas. In total, its water supply operations cover 31 million people. Although this Group is owned 25 per cent by Suez Lyonnaise des Eaux, its recent international successes and its diversification policy give us a clear indication of how the Group could compete with French TNCs other than Lyonnaise. Besides its long history of success in water distribution in Latin America, this group has since the beginning of 1999 won several contracts, and in particular a surprising one in Cuba. Agbar is now responsible for the management of Havana's water, supplying 2.2 million people for the next twenty-five years.[87] Agbar has also become active in the Internet, forming a new affiliate, Agbar Global Market (AGM), to compete in the e-commerce field. AGM combines nine companies that already sell via the Internet, over the phone, or through catalogues. Its chairman, Ricard Fornesa, declared that AGM would have an aggregate profit of E66.1 million in 1999 and further added that he wants AGM to become the e-commerce market leader in Spain.[88] In its forecast for the year 1999, the Group announced a consolidated net profit increase of 11 per cent (£104 million) with an operating income of $2 billion, up 15.5 per cent.[89] These good results are proof enough that this group can be considered as one of the most successful non-French public service TNCs. The group Aquas de Valencia has also had some successful results for 1999 winning two international contracts, one in the Costan Rican capital of San Jose, and the other one for the Venezuelan state of Lara serving a population of 1.5 million people.[90]

87 'Cuba/Infrastructure: Agbar to run Havana's water supply', *Global Water Report*, *Financial Times*, 89 (14 February 2000).

88 'Spain: Agbar on the net', *Global Water Report*, *Financial Times*, 70 (23 April 1999).

89 'Spain: Agbar forecast', *Global Water Report*, *Financial Times*, 63 (9 January 1999).

90 'Venezuela/Privatisation: Spain wins Lara', *Global Water Report*, *Financial Times*, 63 (9 January 1999).

Besides FCC and Aquas de Valencia, both linked with French TNCs, we can find a new group that is slowly emerging in the water sector called Aquas de Bilbao. This firm won at the beginning of 2000 a thirty-year concession to supply potable water and drainage services to the Uruguayan department of Maldonado with a Spanish consortium consisting of Consorcio de Aquas de Bilbao Bizkaia, Iberdrola Energia and the Basque bank BBK.[91] This result is even more impressive, since this consortium was competing in the bidding process against Lyonnaise des Eaux and Vivendi.

Germany

Other potential players can be found in Germany. This is the case, for example, of the Veba Group, now called E.ON AG. This group is a large energy, chemicals, oil, distribution and telecom holding. Energy represented about 20 per cent of its turnover in 1997. Its energy subsidiary PreussenElektra is one of the big three power companies of Germany. The group has been known as E.ON AG, since it has merged with Viag, and now focuses on energy and specialty chemicals. These two leading industrial groups together form the world's largest investor-owned energy service provider, and the world's largest chemicals group. In direct competition with the French TNCs, the merged company wants to become one of the world's premier multi-utilities with its power, natural gas and water activities.[92] The group is also involved in e-commerce through Oneline.[93] However, it should be emphasised that the group is not yet very developed in the water sector. Veba's subsidiary in the water sector is Gelsenwasser, Germany's largest private water supplier. In an interview with the *Global Water Report*, this firm said that it was optimising the operation of its water plants in preparation for future liberalisation of the water market. A spokesman added that it was only a question of time until liberalisation of the energy market would be extended to water.[94] Another firm which is part of the group in the water sector is Hansewasser, which is a joint venture between Gelsenwasser and Stadtwerke Bremen. This firm has won a thirty-year BOT concession to operate the Bremen sewerage system. It is noteworthy to see that the losing bidder was a joint venture between Suez Lyonnaise des Eaux and the Thyssen Group, Norddeutsche Abwasser Bremen Eurawasser.[95] It is evident that when the liberalisation of the water

91 'Uruguay: Maldonado supply', *Global Water Report*, *Financial Times*, 89 (14 February 2000).

92 Press Release of Veba AG (10 February 2000).

93 'Veba's PreussenElektra subsidiary takes majority stake in Oneline', Veba Press Release (17 March 2000).

94 'Germany/Companies: Preparing for the future', *Global Water Report*, *Financial Times*, 90 (2 March 2000).

95 'Germany/Financing: Bremen BOT financing closed', *Global Water Report*, *Financial Times*, 67 (5 March 1999).

markets in Europe takes place, the battle between the French and the German Groups, not forgetting the Spanish and British public services TNCs, will be fierce. Nevertheless, *Veba* has not the international experience the French already have. In order to reinforce its position on the water sector, the Group tried to reach an agreement with Suez Lyonnaise des Eaux.[96] However, it seems that the new E.ON AG Group will not focus at first on the water sector, although the size of the group leads us to consider that it should be seen, with Enron and RWE, as one of the three main challengers to French electricity giant EDF.

The latest news shows that RWE is certainly the main actor to consider alongside the French water TNCs in the water sector. Until now, RWE had focused most on prospecting activities in Central Europe. The Group's initial activity was in electricity. The group then expanded into other sectors, such as waste management and telecommunications. In the late 1990s, some moves indicated that it was also trying to establish itself in the water sector. The first example was the tender it won for the Budapest water supply together with Lyonnaise des Eaux. RWE offered some 30 billion Euro to buy Vivendi Environment, but the deal was finally refused.[97] In September 2000, RWE announced that it had acquired Thames Water for 7.1 billion Euro,[98] which means that RWE is now becoming a serious threat to the French water giants. With this acquisition, thanks to Thames' 24 million overseas customers, RWE Group now has almost 40 million clients. Of course, compared to Vivendi and Suez Lyonnaise des Eaux, the gap is still wide with each of them having slightly more than 100 million customers. Nevertheless, Thames Water will provide the critical mass needed for RWE to compete in the international market in water and other utilities. Thames' know-how in water in Turkey, in Thailand, in Indonesia and in many other parts of the world leads RWE to become one of the major players in the global water market. Indeed, one should recall that through this acquisition, RWE becomes the third biggest group worldwide in water treatment. Moreover, Thames had recently expanded its international basis with a strategic, though minor, contract in the United States with the takeover of E'town. From this point of departure, one could imagine various scenarios for RWE. For many specialists, it is obvious that the water sector in the next two years will further consolidate and concentrate. RWE could be tempted to merge or collaborate with other groups such as Severn Trent, Anglian Water or United Utilities. Moreover, Bouygues said recently that it would like to find a partner for SAUR or to float

96 'Le rachat de Thames Water par RWE sonne le reveil des grands groupes allemands', *Le Temps*, Finance (26 September 2000).

97 'RWE devient no. 3 mondial du tratement des eaux', *Tribune de Genève* (26 September 2000: 15).

98 'RWE devient no. 3 mondial du tratement des eaux', *Tribune de Genève* (26 September 2000: 15).

it on the market in 2001–2. It is thus likely that RWE will become the major competitor of the two French giants.

Italy

Finally in Europe we can also note another successful energy TNC, ENI-Italgas-SNAM. This group is a semi-privatised Italian state energy holding, particularly active at the international level in gas supply through Italgas. ENI-Italgas also operates in water on the Italian market but also at the international level, having acquired, in a joint venture with Enron and SAUR, water services in Mendoza, Argentina.[99]

Outside Europe

Outside Europe, there are only a few cases of public service TNCs. This can be explained by two factors. The first reason for the dominance of European public services TNCs is historical, resulting from past colonial structures. This is further enhanced by the fact that, nowadays, the privatisation of the water sector, combined with the role of development agencies and international organisations, does not favour the emergence of public service TNCs in developing countries. The second reason lies in the state structure and in public services traditions. This is why, for example, we can find few cases of American public services TNCs apart from Enron, to which we shall now turn. Enron and the following examples will in fact show that there is a new challenge for French water TNCs, and that the competitors are not necessarily traditional actors in the water sector. In fact, the following section will highlight the fact that the water sector and infrastructure services are not only interesting to public services TNCs, but also attract new competitors from other sectors. As a result, one can observe an emerging battle between the biggest TNCs worldwide.

Enron is a US company, which pursues an aggressive strategy in electricity, gas and now water. It is very active in Western Europe, Central and Eastern Europe, Asia, Latin America and Canada, as well as in the United States. In the water sector, besides Argentina, the group, through Azurix, is also present in the United States,[100] in Mexico,[101] in Canada, and in the United Kingdom. Indeed, Enron has been presented by most specialists as the main competitor to the two French giants.[102] In 1999, Azurix CEO Rebecca Mark said:

99 'PSIRU Summary Capsule, ENI-Italgas-SNAM', <http://www.btinternet.com/-ipspr/companyinfo/122/cap.htm>.

100 For more details, see 'Canada/Mergers & Acquisitions: Azurix to buy Philip Utilities', *Global Water Report*, *Financial Times*, 69 (1 April 1999).

101 For more details, see 'Mexico: Azurix takes Cancun', *Global Water Report*, *Financial Times*, 64 (23 January 1999).

102 See *Global Water Report*'s 'Watch out: here comes Azurix'.

> We are intending to become the global leader in the water industry. Azurix is looking at three main areas: ownership and management of assets, provision of services to municipalities, including operations and maintenance, and resource management, including water extraction and water management. Those areas encompass a $300 billion revenue market, we think we can be a major, major factor.[103]

Unlike Vivendi and Suez, Azurix and Enron are independent when it comes to strategy. As noted by the Azurix CEO, whilst Azurix remains very much an integral part of the Enron family and multi-utilities are playing an increasingly important role in private sector development, Azurix prefers to stress the autonomous nature of its new venture: 'Azurix is pursuing projects independently of Enron, although there may be some opportunities for joint activity in some fields. There will be some [prospective customers] with water who use a lot of energy and vice-versa, but that's not the basis of our business. A stronger synergy is the transferability of our skill sets from energy to water.'[104] At the beginning of the year 2000, the group revealed yet another strategy which surprised most water specialists. Indeed, whilst the company said it would continue to try the occasional large-city contract, big cities will not constitute Azurix' main objective. This strategic move is explained by the fact that Azurix wishes to narrow its geographic focus and to concentrate more on the service side, which is developing faster than the concession business.[105] This change in direction came at the same time as its announcement that it had entered the world of e-commerce. And indeed, the company has decided to launch an online exchange for buying, selling, storing and transporting water in the western United States. The exchange, as reported in the *Global Water Report*, will be called Water2Water.com in an attempt to bring market forces to bear on water buyers and sellers in California.[106] This service will also post weather reports, crop and commodity prices, reservoir levels, steam flows and other water-related information. The exchange is aimed at water districts, farmers, landowners and other customers. Azurix therefore appears to become a direct competitor to the giant French TNCs, not in the bidding process for concessions, but rather on the technological side. Thus, with the accelerating liberalisation of the water sector we can foresee that the industry will be dominated by just a handful of major players, of which Azurix wishes to take the lead.

Finally, and besides the public services TNCs and Azurix competing over the liberalisation of the water sector, we can also note some important TNCs,

103 'Watch out: here comes Azurix', *Global Water Report*, *Financial Times*, 67 (5 March 1999).

104 'Watch out: here comes Azurix', *Global Water Report*, *Financial Times*, 67 (5 March 1999).

105 'Azurix spells out its strategy: moving away from the cities', *Global Water Report*, *Financial Times*, 89 (14 February 2000).

106 'Azurix changes direction', *Global Water Report*, *Financial Times*, 89 (14 February 2000).

which have no experience in public services, yet are simply attracted by the profits to be made in this sector. For example, in the pre-qualification round for Sofia's water privatisation, one could see among the contenders Japan's Mitsubishi Corporation.[107]

Moreover, some recent signs show that the battle for the water sector is not limited to the distribution of water and sanitation, but extends also into water packaging. This market is seen by many specialist as an emerging one, which will bring into competition groups such as Danone and Nestlé, but also Coca Cola, as well as water TNCs such as Suez Lyonnaise des Eaux and Vivendi. As we have seen above, Vivendi's acquisition of USFilter shows the importance of this new market. Although the European market is still by far the biggest, most competitors focus their strategies on developing countries, where expected growth is well over 10 per cent. For example, Nestlé and Danone are now present in Pakistan, Indonesia and China. The US market is also another important area of growth in packaged water.[108] In other words the dynamics of TNCs in the water sector is still accelerating.

Expansion and integration: the rise of public services TNCs (PSTNCs)

So far in this chapter, we have tried to understand the rapid growth of TNCs. We have also presented the main TNCs in the water sector. In this section, we now would like to analyse this phenomenon of emerging global water TNCs in a more systematic way, by examining their expansion as well as their integration. This process, we claim, will ultimately lead to a few global so-called multi-service TNCs, which will most probably be cartelised.

We have already presented the main TNCs in the water sector. Table 4.8 shows how the twelve biggest ones are spreading globally.

This global expansion of water TNCs can be explained by a combination of two factors, privatisation and the need for investments on the one hand, and environmental pressure and the declared need to consider water as an economic good on the other. Most specialists indeed tend to see these two factors as interrelated. Nevertheless, they stem from two different logics and approaches. The first is to a certain extent ideological as it is grounded in a neo-liberal approach to water and all other utilities: privatisation is seen as a way to better manage infrastructure services, as it introduces commercial principles and competition. Some argue that this approach is ultimately positive for sustainable development, since resources will end up being better managed. Another argument in favour of privatisation pertains to the new

107 'Bulgaria/Privatisation: Sofia prequalification', *Global Water Report*, *Financial Times*, 66 (19 February 1999).

108 For more details on the perspective of this market, see 'Robinets contre bouteilles, la guerre mondiale de l'eau est lancée', *Le Monde* (Thursday 19 November 1998).

Table 4.8 TNCs' global presence in the water sector

Country	*(1)*	*(2)*	*(3)*	*(4)*	*(5)*	*(6)*	*(7)*	*(8)*	*(9)*	*(10)*	*(11)*	*(12)*
Argentina	•	•	•	•					•	•		•
Australia	•	•							•		•	
Belgium	•	•						•				
Bolivia	•										•	
Bulgaria	•		•									
Brazil	•	•							•			
Canada	•		•	•			•				•	
Central African Republic			•									
China	•		•				•		•			
Chile	•	•							•	•		
Colombia	•									•		
Côte d'Ivoire			•									
Cuba	•									•		
Czech Republic	•	•	•						•			
Djibouti		•										
France	•	•	•									
Gabon		•										
Gambia		•										
Germany	•	•			•	•	•	•	•			
Ghana		•										
Guinea			•									
Hong Kong	•											
Hungary	•	•			•							
Indonesia	•	•									•	
Israel	•											
Italy	•	•	•					•				•
Ireland									•			
Lithuania	•	•										
Jordan	•											
Malaysia	•	•									•	
Mali			•									
Mexico	•	•		•						•	•	
Morocco	•									•		
Mozambique			•									
Netherlands							•					
New Zealand		•							•			
Norway	•								•			
Philippines	•										•	
Poland	•	•	•						•			
Portugal	•	•						•		•		
Puerto Rico		•										
Romania	•	•	•									
Russia		•	•									
Senegal			•									
Slovenia	•											
South Africa	•		•									
Spain	•	•	•					•		•		
Sweden									•			

cont.

Table 4.8 (cont.)

Country	*(1)*	*(2)*	*(3)*	*(4)*	*(5)*	*(6)*	*(7)*	*(8)*	*(9)*	*(10)*	*(11)*	*(12)*
Thailand			•									
Trinidad and Tobago								•				
Turkey	•	•										
United Kingdom	•	•	•	•			•	•	•		•	
Uruguay	•	•								•		
United States	•	•		•			•	•	•		•	
Vietnam	•		•									

Notes:
(1) Suez Lyonnaise des Eaux; (2) Vivendi; (3) Saur-Bouygues; (4) Enron; (5) RWE; (6) Veba (now E.ON); (7) Yorkshire Water; (8) Severn Trent; (9) Anglian Water; (10) Aquas de Barcelona; (11) United Utilites; (12) ENI-Italgas-SNAM.
Source: Table created from the PSIRU Database (1 September 1999).

investments needed in the water sector, resulting from both the need for new water infrastructures and from higher environment and sanitary standards. The second logic, as we have already described in previous chapters, stems from the fact that water is becoming more and more precious and is seen by many as an economic good. This reflects an environmental concern, yet does not mean that water should not be available for all. It certainly follows the logic that water is a tradable resource to be allocated by means of full-cost pricing. However, from our point of view, this approach is not necessarily in contradiction with the special nature of water – that is that water should be a public good available for all. This second logic does not, as we know, automatically lead to privatisation. Nevertheless, both logics are often linked together, and as such have been inter-related and ultimately lead to an argument often made, namely that the state should withdraw from the water sector. This argument goes as follows: the abolition of public subsidies and reliance on demand, in order to assess the effective value of supplied water would allow the functioning of flexible markets, reducing waste, pollution and prices as a result of enhanced efficiency (Winpenny, 1994: 9–11). In other words, and as we have argued in Chapter 3, the second logic is used to push for the first, i.e. the growing water scarcity becomes instrumentalised so as to promote water privatisation. In other words, water must be managed by the private sector.

Yet, it appears that such private sector involvement does not necessarily bring about flexible and competitive markets, and in most cases a public monopoly is simply being replaced by a private one. As a result, as with other monopolistic markets, water TNCs will achieve above-average profits. Moreover, such monopolistic tendencies are further enhanced by the fact that water is a less tradable resource than electricity, gas, railways or telecommunications, given the fact that unbundling is almost impossible. Finally, the water sector is characterised by high risks of market failure. With the entry of TNCs in this market, we can thus fear some adverse implications for the public good.

It is thus not surprising to observe a process of market concentration in the water industry. Already, concentration in the water industry is quite impressive, with the control of the world market in the hands of only nine firms, constituting two large groups headed by the French corporations, Suez Lyonnaise des Eaux and Vivendi and three smaller undertakings. (PSPRU, 1996: 8). As we have said elswhere: 'the history of the principal private operators in the global water industry shows that the long ago acquired leadership can be explained in the light of their ability to expand without facing real competition, rather than their early establishment' (Finger and Lobina, 1999: 20).

Water TNCs, like most TNCs, follow a dual strategy, both relying on economies of scale, namely horizontal and vertical integration. Vertical integration can be defined as a strategy aimed at attaining full control of the entire production process up to the finished product. Such vertical direction gives rise to at least two novelties, i.e. the rise in intra-firm trade with the production of intermediate goods on the one hand, and enhanced co-operation in the field of services, and in particular a higher complementary focus on Research and Development (R & D) on the other. The advantage of such a strategy is obvious: it leads to a certain price stability owing to intra-firm trade, as compared to the risks in trading with a third party owing to price fluctuation. Water TNCs, as noted above, have certainly followed this strategy. Yet, they have gone further than that by diversifying into related sectors, from where synergies can be derived. As Table 4.9 shows, water TNCs have expanded into other public services making them prototypes of what we can call 'public service' or 'multi-service' TNCs.

The second strategy consists of horizontal integration, i.e. of extending the range of activities carried out by the firm, supposedly leading to synergies among the different activities. Such integration can be achieved either by means of M&As or by means of joint ventures. The first option reflects an aggressive strategy designed to take control over subsidiaries. On the other hand, joint ventures are seen as a much safer strategy, as this sort of arrangement leads to risk-sharing. As a result of such horizontal expansion, a few TNCs today provide an impressive range of public services in addition to water supply and sewerage, including catering, telecommunications, construction, energy, environmental services, health, housing, security and prisons and transport. Table 4.10 shows the global area in which the main water TNCs are currently active.

In addition, many of these companies display collusive behaviour. By 'collusive behaviour' we mean 'the constitution of combines and cartels, whose members unite the reciprocal market power by co-ordination, in order to impose the desired oligopoly market structure and thus restrain the action of other competitors' (Finger and Lobina, 1999: 28). Table 4.11 shows how the largest public services TNCs are grouping together, leading to just few major groups which basically control the markets all over the world.

Table 4.9 A comparison between Vivendi's and Suez Lyonnaise des Eaux's sectoral integration

Suez Lyonnaise des Eaux	*Vivendi*
Water Water engineering Water treatment engineering Water treatment chemicals 	Water Water engineering Water management Waste water
Energy	Energy Heating
Gas distribution Gas transmission	
Electricity	 Electricity generation
Waste collection and disposal	 Refuse collection
Environmental services	Environmental services
Communications Telecommunications Cable Television Computers	Communications Telecommunications Cable Television Computers Film
Construction Construction-related Highways	Construction
Health services Health care	
Manufacturing	
Prison management	
  Vehicle transportation	Transport Rail
Sports and leisure	Sports and leisure
Consultancy	
Revenue collection	
Parking	
	Cleaning
	Education
	IT
	Property services
Other	Other

Source: Elaborated from the PSIRU database (1 September 1999).

Table 4.10 Presence of public services TNCs worldwide

Parent	*Home Country*	*A1*	*A2*	*A3*	*A4*	*A5*	*A6*	*A7*
AES	USA	•			•			
Aguas de Barcelona	E			•				
Anglian Water	UK		•	•		•	•	•
BET	UK	•	•		•			•
British Gas	UK			•	•		•	•
Camuzzi	I			•				•
Eléctricité de France	F	•	•	•	•		•	•
EDS/General Motors	USA		•	•	•	•	•	•
Endesa	E	•		•				•
Enron	USA		•	•	•			
Gazprom	R						•	•
GdF	F			•	•		•	•
Générale des Eaux	F	•	•	•	•	•	•	•
Houston Industries	USA			•	•			
ISS Servisystem	DK		•	•			•	•
Lyonnaise des Eaux	F	•	•	•	•	•	•	•
National Grid	UK			•	•			
National Power	UK		•		•			•
North West Water (NWW)	UK		•	•	•	•		•
Powergen	UK				•		•	•
Ruhrgas	D						•	•
RWE	D	•			•		•	•
SAUR/Bouygues	F	•	•	•	•	•	•	•
Serco	UK			•	•	•		•
Severn Trent	UK			•				•
Sodexho	F	•	•	•	•		•	•
Southern Company	USA		•		•		•	
Thames Water	UK	•		•	•	•	•	•
Tractabel	B		•	•	•		•	•
Veba	D	•	•	•	•		•	•
Viag	D		•				•	•
Welsh Water	UK		•		•	•	•	•
WMX	USA		•	•	•	•		•

Notes:
Regions: A1 = Africa; A2 = North America; A3 = South and Central America; A4 = Asia; A5 = Australasia; A6 = Central and Eastern Europe; A7 = Western Europe.
Home countries: B = Belgium; D = Germany; DK = Denmark; E = Spain; F = France; I = Italy; R = Russia; UK = United Kingdom; USA = United States.
Source: Finger and Lobina (1999).

Collusion is generally done through joint ventures, but collusive agreements are also entered into between groups and enterprises through large consortiums, especially at times of bidding.

But the most noteworthy phenomenon we can observe in the case of water TNCs is 'multi-utilities' or 'multi-services'. As we will see in later chapters, Suez Lyonnaise des Eaux and Vivendi have developed this multi-utilities

Table 4.11 Collaboration and joint ventures involving water TNCs

Company Partner	*Country*	*C1*	*C2*	*C3*	*C4*	*C5*	*C6*	*C7*	*C8*	*C9*	*C10*
Aguas de Barcelona	E		•	•	•	•					
Alcatel	F				•						
Anglian Water	UK	•			•	•					
AT&T	USA				•						
Bechtel	USA							•			
Canal de Isabel II	E	•									
Eléctricité de France	F				•	•		•			
Endesa	E	•		•		•		•			
GdF	F	•									
Générale des Eaux	F	•	•			•				•	
Hidrocantabrico	E	•			•						
Hydro-Quebec	C							•			
Italgas	I				•						
Italomobiliare	I					•					
Lurgi Bamag	D										•
Lyonnaise des Eaux	F	•	•		•						
Mitsubishi	J					•					
Pirelli	I				•						
RWE	D				•						
Thames Water	UK				•						
Thyssen	D					•					
Total	F					•					
Tractabel	B										•
Wessex Water	UK	•									

Notes:
Companies: C1 = Aguas de Barcelona; C2 = Anglian Water; C3 = Canal de Isabel II; C4 = Générale des Eaux; C5 = Lyonnaise des Eaux; C6 = North West Water (NWW); C7 = SAUR/Bouygues; C8 = Severn Trent; C9 = Thames Water; C10 = WMX Technologies.
Countries: B = Belgium; C = Canada; D = Germany; E = Spain; F = France; I = Italy; J = Japan; UK = United Kingdom; USA = United States.
Source: Finger and Lobina (1999).

strategy so as to become world leaders in the provisions of public services to local communities. This strategy, leading to the creation of horizontal industrial empires, is designed to allow future growth along several lines of business, as well as to offer an integrated package of public services on the occasion of competitive bids and beyond (PSPRU, 1996: 1–7). Table 4.12 shows the various sectors in which these multi-services TNCs are currently active.

There are numerous examples of public services TNCs adopting a multi-utilities strategy. Vivendi, for example, says in its 1998 *Annual Report*:

> Vivendi is targeting 'multi-utility' contracts, where the distribution of water, gas, and/or electricity for a customer is covered by a single

Table 4.12 Main TNCs operating across a number of public services

Company	*Main Business*	*Sector* / *Home Country*	*Ca*	*Cl*	*Co*	*Cp*	*Cs*	*Ed*	*En*	*Ev*	*He*	*Hg*	*M*	*P*	*SP*	*T*	*W*
Aguas de Barcelona	Water	E					•		•		•						•
BET	Cleaning	UK	•	•			•							•	•		
Eléctricité de France	Energy	F							•	•							•
Endesa	Energy	E			•				•	•							•
Enron	Energy	USA							•				•				
Générale des Eaux	Water	F	•		•		•	•	•	•	•	•		•		•	•
Lyonnaise des Eaux	Water	F	•		•		•		•	•	•		•		•		•
North West Water	Water	UK	•	•	•	•											•
P&O	Shipping	UK	•	•			•							•	•		•
RWE	Energy	D			•		•		•	•							
SAUR/Bouygues	Water	F			•		•		•	•							•
Serco	Defence	UK				•	•	•		•		•		•	•		
Severn Trent	Water	UK								•							•
Sodexho	Catering	F	•												•		
Southern Company	Energy	USA			•				•								
Thames Water	Water	UK								•			•				•
Tractabel	Energy	B	•	•	•												
Veba	Energy	D			•				•								
Viag	Energy	D			•				•								
Welsh Water	Water	UK						•									•
WMX	Environment	USA							•	•							•

Notes:
Sectors: Ca = Catering; Cl = Cleaning; Co = Communications; Cp = Computers; Cs = Construction; Ed = Education; En = Energy; Ev = Environmental; He = Health; Hg = Housing; M = Manufacturing; P = Property; SP = Security and Prisons; T = Transport; W = Water.
Home Countries: B = Belgium; D = Germany; DK = Denmark; E = Spain; F = France; I = Italy; UK = United Kingdom; USA = United States.
Source: Finger and Lobina (1999).

contract. The multimedia concept is growing in popularity in Europe, and taking hold in some emerging markets. This is the case in Gabon, where Générale des Eaux was awarded a 20-year concession to produce and distribute drinking water and electricity to the country's 1.5 million people. This innovative approach opens new possibilities for agreements with local authorities wanting to outsource several utilities – waste, water, energy, for example – to a single operator'.[109]

The same strategy has also been used by the Bouygues Group in Ivory Coast where, through its subsidiary Saur which had entered the water market, Bouygues acquired a few years later a contract for the electricity network. This strategy is also currently used by Vivendi in the United States, where Vivendi made a bid for the US solid waste company Superior in June 1998. With this acquisition the company claims that its waste management subsidiary, Onyx will 'have the means of building a North American base with critical mass to develop all its waste activities, in line with its worldwide strategy'.[110] This and other acquisitions (such as the non-nuclear power generating plants of GPU and USFilter in April 1999) mean that Vivendi has now built a US position in water, energy and waste management. The company's strategic goal is to break into the US market, offering multiple environmental services to all its customers, i.e. municipal authorities, manufacturers and consumers.

Conclusion

In conclusion, we would like to make three remarks, concerning the many shortcomings of water privatisation. The first shortcoming pertains to the *concentration* process among TNCs in general, and specially public services TNCs. Indeed, one can foresee, in the area of public services, concentrations similar to the ones in the car industry or the oil sector, for example.[111] In this chapter, we have examined the important firms involved the water sector. With the liberalisation of the European water market, and more generally the world water market, the process of concentration and take-overs seems inevitable and it is likely that only three or four main groups will dominate and control the world's water market by 2005 or so. Among these, we will almost certainly find Suez Lyonnaise des Eaux, Vivendi, RWE and maybe Enron. This also means that the initial impetus given by the World Bank in reforming

109 Vivendi, 'The year in brief, 1998, Utilities, Construction and Property, Communications, Vivendi', *Annual Report, 1998*: 24.

110 Vivendi Press Release (14 June 1999).

111 In the car industry, the top five companies control over 50 per cent of the market. In the oil industry, the top five companies control over 40 per cent of the market (UNCTAD, 2000: 128).

the water sector, and more generally infrastructures, has not brought about the desired the effects. Rather the contrary is the case: as a result of World Bank privatisation we will have neither a market nor competition, but rather monopolies, oligopolies and cartels. What is worse, the World Bank may simply have been instrumental in helping some French and other public services TNCs to acquire global monopoly power.

And this lack of competition also has another important consequence: indeed, although private sector participation has introduced new investments which were indeed badly needed for reforming the water sector, the monopolistic character of the market and the high concentration process described above will not lead to water becoming treated as an economic good, which was originally intended in order to make economic actors behave more sustainably. Rather, water privatisation will simply make people pay (more) for water, without really changing their behaviour. In other words, privatisation has not brought about competition at the price level and has not altered the perhaps natural monopolistic character of the water market.

Our final point pertains to the fact that we can no longer just talk about water TNCs. Rather, we must now talk about public services TNCs or multi-utilities. Yet, the concept of public services TNCs remains full of contradictions. If the purpose of privatising water, at least in the eyes of the World Bank, was to remove the state from operations, the fact of the matter is that some (minimal) 'public services obligations' will almost certainly remain. This in turn means that the state – or some other public entity, for that matter – will continue to be involved in the water sector. Not to mention the monopolistic nature of the sector, which will also call for state intervention. Paradoxically, one might well end up with more state presence in the water sector after privatisation than this was the case before.

Chapters 5 and 6 will show examples of water privatisation in the developing countries, transitional economies, and the North. Yet, in all these cases we will be able to observe again the above three shortcomings – concentration, lack of competition and declining state presence.

Chapter 5

Water resources management in developing countries and Eastern Europe

> Standing at the threshold of the 21st century, the urban water supply and sanitation sector in developing countries faces major challenges. Continuing population growth and rising standards of living mean that safe water and adequate sanitation is to be supplied to ever-increasing numbers of households in ever-increasing quantities.
>
> (Blokland *et al.*, 1999: 3)

Chapters 5 and 6 will present case studies from the South, the East and the North. Through these case studies, we seek to illustrate the main points made so far, namely the spreading of the new approach to water resources management, the central role played by the World Bank and, above all, the rise of water or public services TNCs. The purpose of these two chapters is twofold: on the one hand, they will illustrate how each case is different, and that it is therefore difficult to apply one single model to water resources management, as is done by the World Bank. On the other hand, the presentation of these cases will serve as empirical material for developing, in Chapter 7, recommendations for water resources management in the future.

One might question our distinction between developing and Eastern European countries on the one hand and industrialised countries on the other. Indeed, the public services TNCs active in the water sector are the same in the North and in the South. Public sector reform, deregulation and privatisation are also not fundamentally different in rich and in poor countries. And we think that, in the end, there should not be a big difference in the way water resources are managed in the various parts of the world, especially when it comes to regulating the private operators. Nevertheless, there still remain, in our view, three main differences between the South and the North, and which warrant a different treatment, namely the state, markets and the strategies of the public services TNCs:

- Be it in developing or in Eastern European countries, the *administrative structure* and the *judicial system* are usually very weak in terms of both human and financial capacities. Privatisation and decentralisation (of water resources management) therefore often need to be preceded by substantial capacity building and institutional development in order to be able to handle the increasingly autonomised operators and TNCs, as well as to satisfy the users.

- Similarly, *markets*, particularly in the developing countries, are generally still underdeveloped, except for cities and urbanised or otherwise densely populated areas. Especially in developing countries water production, distribution and sewerage infrastructures often do not exist or are in very bad condition, creating substantial investment needs before any revenue can even be generated.
- Both observations lead water TNCs to behave very differently in the North and in the South, which does of course not mean that their earnings per user are less in the developing countries than in industrialised countries. Actually, quite the contrary appears to be the case, which ultimately is the explanation why water and public services TNCs will undertake all the hardships to be present in the developing world. In short, the main difference between the North and the South – as well as the East for that matter – stems from the fact that in the South and the East TNCs uses *World Bank sector reform projects* in order to enter the markets.

In this chapter, we seek to cover all the main regions of the world, i.e. Africa, Asia, Central and Latin America and Eastern Europe, showing each region's particular problems, and highlighting in each case how water privatisation is related to these problems. For example, Africa is faced with growing poverty and lack of investment in the water sector causing low, if any, access to water and sanitation facilities to the poor. We will see in our case study (Ivory Coast) in particular what access to water means for the urban poor. In the case of Asia, rapid industrialisation and intensive agriculture raises in particular the question of how environmental concerns can or cannot be addressed through water privatisation. Here, we will in particular examine the case of the Philippines. South and Central America, in turn, are particularly confronted with decentralisation, an issue we will examine in the case of El Salvador. This chapter will thus examine the consequences of private sector participation in water resources management with regard to three main challenges – investment, environment and decentralisation. In conclusion, we will also discuss water privatisation in Eastern Europe, which not only combines all three themes, but moreover adds an explicitly ideological dimension.

Africa: the investment challenge

Africa's main problem is not population growth – there will be about 1 billion people living on the continent in 2025 – but poverty and rapid urbanisation.

Indeed, Africa is still the least urbanised continent of the world, yet is currently experiencing the world's most rapid rate of urbanisation, approximately 5 per cent per year. According to Damhaug, there will be a population increase of over 370 per cent in African cities, from an average of 138 million in 1990 to 500 million in 2020 (Damhaug, quoted in Sharma *et al.*, 1996: 18). While some African countries have more than enough water on a *per capita* basis, population growth, urbanisation and economic development create demand exceeding supply for most countries. In fact, in 2000, about 300 million Africans lived in a 'water-scarce environment'. In 1990, eight countries were suffering from water stress or scarcity, and by 2025, the number of countries experiencing water stress will rise to eighteen, affecting some 600 million people (Sharma *et al.*, 1996: 1). In 1991, for example, only 44.9 per cent of the African population had access to potable water, with considerable variation in access among countries. This must be compared to 68 per cent in East Asia and the Pacific, and to 76 per cent in Latin America and the Caribbean. The low rate of access in sub-Saharan Africa is due to a number of factors, including insufficient network development, pricing problems, low availability of water and competition between alternative uses.

Figures 5.1 and 5.2 illustrate population growth with percentages of populations served and unserved with water and sanitation services. They show that in 2020 approximately 20 per cent of Africa's population will have access to water supply services and approximately 10 per cent to sanitation services.

Currently, about 265 million people in Africa do not have access to potable water and 344 million do not have adequate sanitation facilities (Deveaux, 1998: 75). This is particularly the case for rural and peri-urban areas. Indeed, according to World Bank figures, about 65 per cent of the rural population and 25 per cent of the urban population are without adequate water and, respectively, 73 per cent and 43 per cent without proper sanitation (Sharma *et al.*, 1996: 2). In the years to come these figures can only increase. Investment

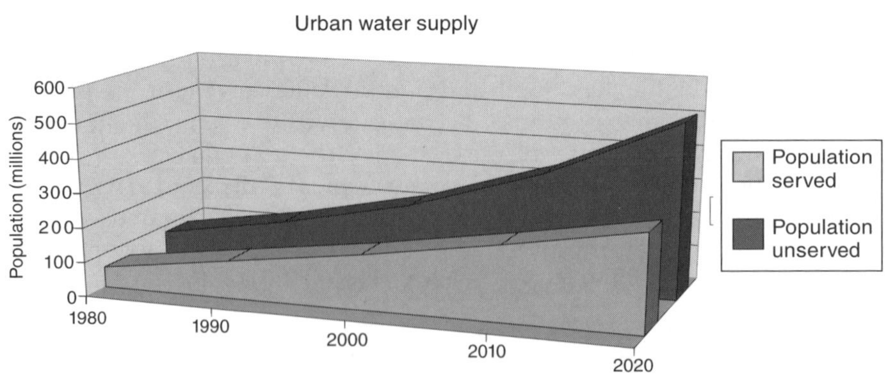

Figure 5.1 Served/unserved population in urban and rural areas, water supply, 1980–2020. *Source*: Sharma et *al.* (1996: Figure 2.11).

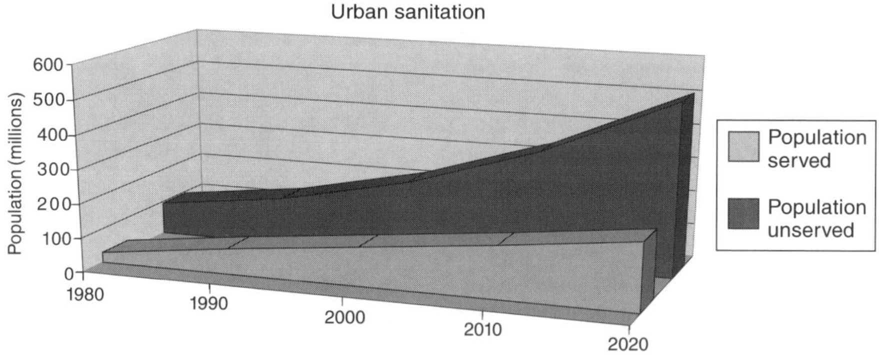

Figure 5.2 Served/unserved population in urban and rural areas, sanitation, 1980–2020. *Source*: Sharma *et al.* (1996: Figure 2.11).

in water production and sanitation could indeed substantially minimise disease. While coverage is expanding relatively rapidly, it is not keeping pace with population growth and urbanisation, and growing demand will in turn increase the cost of water. Indeed, since most accessible surface and groundwater have already been exploited, new water sources will have to be developed, often at a distance and at additional costs. Water supply projects financed by the World Bank in Africa show that the unit cost of water will double if not triple in the near future (Sharma *et al.*, 1996: 22).

Yet, poverty remains among the biggest problems in Africa. Unlike most other continents, poverty has not been decreasing, but has even been exacerbated because of population growth and conflicts. In sub-Saharan Africa, the number of poor people increased from an already high 217 million to 291 million between 1987 and 1998, leaving almost half the residents in poverty by World Bank standards. Figure 5.3 compares the changes in the number of people living on less than 1 US$ per day between 1997 and 1998.

At the same time, governments in sub-Saharan Africa are under considerable financial pressure. Not surprisingly, many of them would like to withdraw from owning and operating infrastructure services. Consequently, some countries have turned to the private sector for financing the badly needed investments, as well as for operating these infrastructures. About thirty African countries, members of UADE (*Union Africaine pour le Dévelopement de l'Eau*), have therefore decided to hand the management of parts of their water sector to private operators. According to Deveaux (1998), such 'privatisation' has proven efficient in such countries as Ivory Coast, Gabon, Senegal and Guinea. However, one should note that since 1984 over 550 infrastructure enterprises have been privatised in some eighty-six countries, and over 580 green field projects are underway in some eighty-two countries. Of these 1161 private infrastructure undertakings, sub-Saharan Africa has seen only eighty. This represents just 7 per cent of the world's total privat-

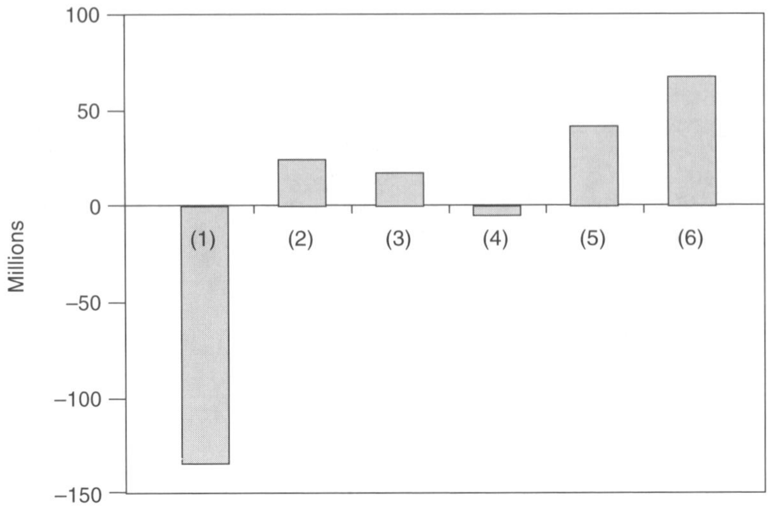

Figure 5.3 Change in the number of people living on less than $1 a day, 1997–8.
Source: World Bank (2000: 4).

Note:
(1) East Asia and Pacific; (2) Europe and Central Asia; (3) Latin America and the Caribbean; (4) Middle East and North Africa; (5) South Asia; (6) sub-Saharan Africa.

isation projects. In the wastewater sector, there have been, according to the World Bank Private Infrastructure Project Database, less than ten projects in sub-Saharan Africa since 1984 (Kerf and Smith, 1996: 6). Yet, the World Bank sets great store by 'the promise of infrastructure privatisation'. We thus have chosen to present the example of the Ivory Coast in order to examine how privatisation affects the access of the poor, and in particular whether privatisation provides for the investments to increase coverage, while providing affordable prices for the urban and the peri-urban poor.

Case study: Ivory Coast

Most water specialists refer to the Ivory Coast as a successful model in water resources management for sub-Saharan Africa when it comes to private sector participation.[1] This can at least in part be explained by the fact that this is not a recent phenomenon. In this section, we will thus briefly present the history of private sector involvement in Ivory Coast, as well as the current institutional arrangements, and we will confront water privatisation with the double challenge of investment and serving the urban poor.

1 Abidjan, for example, is actually under the UNEP/UNHCS' Managing for African Cities programme, where it is used as one of the seven demonstration cities for the African continent.

History of private sector involvement in Ivory Coast water provision

The entry of private sector participation in the water and sanitation sector in Ivory Coast is in fact a very interesting example, as this dates back to as early as 1959. In this year, SAUR, following an international auction, became Abidjan's water supplier. According to Lavigne, this can be explained by the fact that 'the decision to confer part of a public service to a private company belongs to the President and his original ideas about liberalism and help by foreign capital. It also belongs to the relations the chosen company had with the leaders of the country' (Lavigne, 1995). However, it was not until 1974, following the *Programme National de l'Hydraulique Humaine*, that Société de distribution d'eau Côte d'Ivoire (SODECI, the new name of the new company) became the sole operator in the water sector,[2] and was now in charge of the management and maintenance of the whole water supply network in Ivory Coast. SODECI had obtained a fifteen-year lease contract from the government for the whole territory, except for Abidjan, which was under a specific concession contract.

Yet, the quite positive developments of the 1970s were followed by a general economic crisis in the early 1980s, negatively affecting Ivory Coast's development and SODECI's operations. This situation caused a decline in investments and put an end to network expansion. Government had to step in to subsidise small users' water bills. Despite such efforts, 90 per cent of SODECI's consumers were (and still are) facing economic constraints and consume(d) less water than previously. Indeed, the national average of water consumption was 322 m^3 per year in 1987, while it was 886 m^3 per year in 1973. As a result, despite a social programme, an important part of the population either stopped the contract with the company or was unable to pay the monthly bills. In 1987, a new agreement was signed in an attempt to overcome some of these difficulties and SODECI was awarded a twenty-year concession by the Water Board (called 'Direction de l'Eau').

This shows that agreements between government and operator cannot always be sustained, and often must be adapted to the evolving situation. Indeed, the government's and the private operator's objectives are generally different, and the institutional arrangements between these two appears to be of paramount importance. Let us thus examine the current institutional arrangements in more detail.

Current institutional arrangements

The *Haut Commissariat à l'Hydraulique* (HCH) is responsible for water resources management and sector regulation. To meet the objectives of the

2 Indeed, until 1974, two enterprises, one public (Energie Electrique de Côte d'Ivoire, EECI) and the other private (SODECI) shared responsibility for the water sector.

national human hydraulics policy, another authority had been designated in order to establish the administrative and technical structures necessary for the realisation of the programme. After a few institutional changes, the Water Board (Direction de l'Eau) was created in April 1984. Located within the Ministry of Equipment, Transports and Telecommunications, this Water Board is responsible for technical planning, design, conception and control of urban sewerage and drainage facilities, as well as for the control of water quality.

As an operator, SODECI is responsible for the treatment, the supply and the distribution of water. It is also in charge of the maintenance of the network, as well as of the extension of the current network to new providers. These new investments are financed through a Water Development Fund, thanks to a tax on the sales price of water (Lavigne, 1999: 50).

The HCH had established a general water policy for Ivory Coast structured along societal, economic and policy, as well as institutional objectives, broadly following the Dublin Principles (see Chapter 2). Among these principles, let us consider in particular the relation between the regulator and the operator. Basically, HCH has the power to force the private operator to maintain, develop and guarantee access to the network, especially in regards to two societal objectives, which are:

- to allow access to potable water at all the levels of society, particularly for the disadvantaged people, by putting an affordable price on water
- at an individual level, to provide each user with a sufficient amount of water for his/her needs.

In this case study we can unfortunately look only at Abidjan. Yet a systematic survey of water provision across the entire country would certainly be much more valuable, especially when it comes to the presumed discrepancy between urban and rural areas. Since Abidjan is under a concession contract, it is the operator, which is in charge of the investments. The 1987 contracts gave SODECI the responsibility for submitting investment plans for renewals, extensions and social connections (e.g. Kerf and Smith, 1996: 31). Let us see to what extent these are respected and to what extent such private operation actually benefits the urban poor.

Investments and reaching the urban poor

Abidjan is a city which is surrounded by important water resources reserves and water demand can easily be met. This, however, might be quite untypical for Africa, even though the Abidjan water supply system has been seen, until now, as a model of success for the African continent, or at least for the rest of West Africa. Indeed, if we look at the recent financial data available, as well as at efficiency and production/distribution results, private sector participation has certainly been positive (Table 5.1).

Table 5.1 SODECI's financial data in million CFAF and $US, 1993–6

	1993	*1994*	*1995*	*1996*	*Variations 1996/1995*
Turnover		CFAF22,586 ($37.64)	CFAF24,908 ($41.51)	CFAF27,883 ($46.47)	11.94
Total charges		CFAF21,558 ($35.93)	CFAF23,613 ($39.35)	CFAF26,011 ($43.35)	10.16
Net profit		CFAF1028 ($1.71)	CFAF1295 ($2.16)	CFAF1872 ($3.12)	44.56
Frozen assets	CFAF10,204 ($17)	CFAF12,451 ($20.75)	CFAF13,446 ($22.41)	CFAF13,736 ($22.89)	2.16
Long-term debts	CFAF6149 ($10.25)	CFAF6251 ($10.42)	CFAF6750 ($11.25)	CFAF6656 ($11.09)	−1.39

Source: SODECI (1996).

Such financial data, however, do not reveal whether the company is investing and expanding its services. Indeed, increased efficiency and technology can on their own increase the company's net profit. In terms of efficiency, the company has been very successful – since 1996, for example, operating efficiency in urban areas has been excellent with a cost recovery rate of 85 per cent, a 13 per cent rate of unaccounted-for-water and a high collection rate of 98 per cent for private customers (SODECI, 1996). Comparatively, SODECI's productivity appears to be twice as high as that of any water utility in West Africa (Haller, 1998: 26). In terms of production/distribution, Table 5.2 indicates that the water network has only slightly increased.

Indeed, by 1995 about 82 per cent of Abidjan's population had access to adequate water supply facility in urban areas, compared with less than 30 per cent in 1974.[3] However, these figures do not show the 'informal sector'. There are indeed major differences between the numerous districts in Abidjan and most of the poor areas in the urban periphery are not connected to the water piped system. Thus, with an average of ten persons per household, it appears that only half of the urban population (3.45 million) is receiving piped water directly at home. In Abidjan, the main part of the other half has access to safe water by purchasing it from street vendors. As Laurence Haller says,

> the reasons why poorest urban dwellers are often driven to rely on water vendors are: (1) a lack of convenient standpipes, itself often a consequence of government policy not to provide infrastructures in 'illegal'

3 UNDP–*World Bank Water and Sanitation Programme, Workplan* (January–December 1998), quoted in Haller (1998: 27).

Table 5.2 Potable water production and distribution, 1973–96

Facility	*1973*	*1980*	*1993*	*1995*	*1996*
Number of localities equipped with piped water	38	139	409	410	411
Installed network length (km)	922	6000	10,000	10,699	
Consumers number	40,071	130,478	275,152	318,502	344,855
Sold water volume (thousand m^3)	35,528	72,065	88,659	93,344	100,064
Water production (thousand m^3)	43,326	90,096	102,810	110,621	118,147

Source: SODECI (1996).

> unplanned squatter areas, and (2) a lack of money. Poor urban dwellers have often not the means to pay the deposit of 28 000 CFAF to SODECI or they are simply not able to pay monthly bills . . . They can afford to pay water on a day-to-day basis, but not on a monthly basis, even if it is more expensive.
>
> (Haller, 1998: 29)

Also, the objectives set up by the HCH and the Water Board are not respected. Indeed, it appears that SODECI has not the capacity to increase the existing system. Indeed, as pointed out by Laurence Haller

> the pricing policy of SODECI is aiming at a long-term self-financing of the company, through the establishment of two funds. Cost recovery is high enough to provide capital for maintaining existing systems, but remains too low for developing new infrastructures. Despite a slight decrease since 1995 of the debt amount, SODECI is still dependant on external financing to extend its water supply system.
>
> (Haller, 1998: 35)

In other words, the private sector, although more efficient in terms of financial results, has not been able to overcome the problems which existed prior to privatisation. Access of the urban poor and the investments needed for providing such access apparently cannot be guaranteed by market logic alone, either because there is no market, or because the profits of the private operators preclude access and quality. Even an international financial organisation, such as the African Development Bank (ADB), seems to have come to this conclusion:

> Poor infrastructure remains the single most important obstacle to development, and water supply and sanitation projects are among the most likely infrastructure activities to help in the process of pro-poor growth. Where the private sector investment is not forthcoming or where the market fails, the ADB recommends that the state provide vital services, especially for the rural and urban poor. State involvement can have significant results if it is supported by increased local community participation and assistance from the international community.[4]

One may therefore ask why privatisation was promoted to begin with. Yet, even where markets fail, private sector investments may still be needed. However, these private operators needed to be regulated so that they not only provide the agreed-upon universal coverage, but moreover undertake the investments to which they have committed.

Also, we have presented here only the case of Abidjan, but one should not forget the entire country. If the private sector – i.e. water TNCs – is not capable of providing access to the urban poor, how can it then provide satisfactory services to the rural population? In other words, even with a coherent national water policy and a strong institutional setup for regulation, financial support necessary infrastructure investments will continue to be necessary. Hence the importance of World Bank or donor financing.

However, the problem here is that such financing is contracted between the government and the private operators, even though the government generally does not have the capacity to professionally monitor, and even less to enforce, such contracts. Not to mention the fact that setting up regulators in parallel to privatisation is generally not financed by the World Bank. In short, privatising without simultaneously, or even previously, setting up regulators and strengthening government capacity in general will in most cases be a recipe for disaster, leading to undue enrichment by the private operators. Knowing this, one could also ask whether it would not be the World Bank's responsibility to make sure that these private operators – i.e. the few public services TNCs which generally get these contracts – are properly regulated, if need be at a global level? For the time being, things are rather the other way around: by promoting privatisation, the World Bank helps water and other public services TNCs to get access to lucrative markets – i.e. in general in densely populated urban areas – without simultaneously creating the conditions for these TNCs to be regulated. What is worse, these TNCs most generally operate at a local level, which means that the national level, where such regulation should occur, is often not even interested.

4 'Africa: Private and Public Roles in Water Supply', *Water and Sanitation Weekly*, Water Supply and Sanitation Centre and IRC International Water and Sanitation Centre, 34–35 (6 September 1999) <http://www.wsscc.org/source>.

Asia: the environmental challenge

Asia's development presents a formidable challenge to water resources, in particular because of growing environmental degradation, population increase and urbanisation. In 1990, for example, some 2.9 billion people (55 per cent of the total world's total population) were living in East and South Asia, and the UN forecasts that this population will increase to about 4.2 billion in 2025. On the other hand, Asian countries occupy only about 16 per cent of the world's total land surface (Frederiksen *et al.*, 1993: xiii). Moreover, in 1990 the urban population already accounted for 24 per cent of the total population in East and South Asia, and this figure will increase, according to UN forecasts, to about 50 per cent by 2025. Therefore, we can anticipate large increases in water demand for urban supplies and industries, which will in parallel increase the problems of urban and industrial waste water disposal. In short, growing industrial and agricultural production, combined with current demographic trends, will exacerbate environmental problems, both in the water sector and elsewhere.

Indeed, pressures on water resources are numerous and the natural equilibrium of water flows is already significantly affected by agriculture and industry. Dam building is probably the most obvious example, where water production and environmental protection clash. Moreover, projects such as the Three Gorges dam in China or the Sardar Sarovar dam in India have shown that Asia is probably the region where dam building has become most controversial in recent times.

Besides dam building, Asia is also a continent characterised by an important agricultural sector and irrigation is by far the largest water user and consumer in virtually all Asian countries. Asia is also the continent of the 'Green revolution', involving fertilisers and pesticides, high-yielding seeds and irrigation. The environmental effects of such high-yield agriculture are the pollution of rivers, lakes, coastal waters and groundwaters. Indeed, according to the World Bank, polluted waters and related impacts pose increasing environmental and health problems, and this process of environmental degradation in Asia still seems to be in its infancy.

Indeed, population pressure leads to opening up additional land for cultivation, which in turn means the expansion of agriculture into marginal and fragile lands, as well as the invasion of forest areas by unsustainable agricultural practices. As a result, as Table 5.3 shows, agriculture is by far the biggest user of water. In more than half of the Asian countries, agriculture accounts for 90 per cent or more of total water use.

Most water is therefore basically used for irrigation, as in Asian agriculture around 30 per cent of croplands in average are under irrigation. The most striking examples here are India and China. Indeed, in China, the total area under irrigation has grown from about 20 million ha in 1949 to 46.6 million ha at present, whereas in India, the total area under irrigation expanded from

Table 5.3 World Bank-supported water resources development projects, by country and type, 1947–90

Country	*Multi-purpose*	*I&D*	*FCD*	*WS&S*	*Hydro-power*	*Inland Navigation*	*Inland Fisheries*	*Total*
Bangladesh	–	12	10	5	–	2	2	31
China	–	2	1	1	3	–	1	8
India	2	42	1	13	6	–	1	65
Indonesia	1	22	3	7	2	–	–	35
Korea	1	3	–	4	–	–	–	8
Laos	–	–	–	–	–	–	–	1
Malaysia	–	4	–	4	1	–	–	9
Myanmar	1	2	3	–	–	1	–	7
Nepal	–	10	–	4	4	–	–	18
Pakistan	4	14	3	5	–	–	–	26
Philippines	3	15	–	12	3	–	–	33
Sri Lanka	1	8	1	3	1	–	–	14
Thailand	–	10	–	2	5	1	–	18
Vietnam	–	1	–	–	–	–	–	1
Total	13	155	22	60	26	4	4	274

Note: I&D (Irrigation and development); FCD (Flood control and drainage); WS&S (Water supply and sanitation).

Source: Frederiksen *et al.* (1993: 116).

21 million ha in 1950 to 56 million ha in 1989 (Frederiksen *et al.*, 1993: 12). Despite that, several Asian countries have failed to keep growth in food production in line with their population growth over the past decades. It is therefore not surprising that the World Bank, as well as the ADB, want to pursue the development of water, despite the environmental problems such development causes. As the Bank says in a 1993 paper:

> The environmental costs of large projects have rightly received attention. But the environmental costs of not undertaking such projects can also be substantial. The question is not whether a project is big or small, but whether on balance, it is good or bad . . . Given the size of Asia's river systems, the populations concerned, and the costs and dislocations that would result from large-scale reallocation from agriculture, further massive water resource development programs appear inevitable. For example: The only way that dry season flows in the Ganges can be augmented to serve huge populations and contain further sea water intrusion in the delta is through large dams to capture surplus monsoon runoff; the only way that the long-run needs of a population of more than 300 million living on the North of China Plain can be satisfied without catastrophic overuse of Yellow River or groundwater supplies is through transfers from the Yangtze; and the only way that Bangkok can be supplied and further land subsistence avoided while accommodating

> agriculture in the Chao Phraya Basin will probably be through water imports, possibly from Mekong. The underlying dynamics of population growth and economic development are inevitable, and, therefore, major water development programs are inevitable.
>
> (Frederiksen *et al.*, 1993: 9)

Table 5.3 shows Bank-supported water resources development projects between 1947 and 1990.

About 69 per cent of the projects have direct irrigation benefits, which shows the importance given to food production. Agriculture, together with dam building, is therefore the main factor affecting the quantity and quality of water resources available. The current trends in most Asian countries are worrisome, and as a result of this evolution, there will be increasing competition for water use between sectors (domestic, industrial, agriculture) and perhaps to a certain extent between countries. In China for example, competition for water use has already overreached its limit, in particular on the North China Plain, where the surface water is fully developed and the groundwater overdeveloped in extensive areas (Frederiksen *et al.*, 1993: 15). Competition between countries is now also a reality – with, for example, the decision of India to divert part of the Ganges affecting Bangladesh water resources.

In short, in the water sector, Asia appears above all to be confronted with environmental problems. Our case study will therefore focus on the relationship between privatisation and the environment. However, the Bank had not always recommended privatisation. As late as 1993, it still had a very ambiguous position in this respect, when it said:

> Private development of captive sources, rather than being an objective to be desired, is usually a response to a poor public service. Where such development is unregulated, this can have undesirable effects, for instance overdraw of groundwater leading to land subsistence and/or saline intrusion (e.g., in Bangkok and other cities). Rather than the promotion of such forms of privatisation, major emphasis must be placed on regulation and the provision of an improved and reliable public service.
>
> (Frederiksen *et al.*, 1993: 43)

The question therefore here is how water privatisation can be reconciled with environmental sustainability.

Case study: The Philippines

According to the World Resources Institute (WRI), the Philippines, compared to other Asian countries, already has a relatively high rate of access to safe drinking water and sanitation. Indeed, in urban areas, access to safe drinking water is said to be 93 per cent and 77 per cent for sanitation. In rural areas, access is still relatively high, with 77 per cent for safe drinking water and 67 per cent for sanitation.[5] However, access is one thing, water quality another, and the overall management of water resources a third. In this section, we will again present the history of private sector participation in the Philippines, discuss the corresponding institutional arrangements and analyse in particular the relationship between water privatisation and environmental protection.

History of private sector participation

Water privatisation in the Philippines so far seems to be limited to Metro Manila's Water and Sewerage System (MWSS). Apparently, the decision to privatise MWSS was taken by the government, because of MWSS's 'failure under public management to provide adequate water supply and sewerage services to the largest urban centre in the Philippines and the desire to end government subsidies to its operations' (David, 1999: 84). Petrella reports that 40 per cent of Manila's population did not receive water on a regular basis, that 50 per cent of the water was lost to leaks, that sewerage was not working and that accessible water was largely polluted (Petrella, 1999: 21).

Therefore, a local union reports,

> in November 1995, for a fee of US$ 6.2 million the IFC (International Finance Corporation), which is an affiliate bank of the World Bank, was commissioned by MWSS as foreign consultant to draft a privatisation strategy for MWSS. Out of the total fees paid to IFC as consultant, $2.2 million is a success bonus which will be paid directly by the winning bidder, while the other $3 million in retainer's fee and hotel expenses of IFC officers will be reimbursed to the MWSS also by the winning bidder. Incidentally, the IFC also engineered the privatisation of Aquas Argentinas in 1992 and had a 5% equity share in the same entreprise.[6]

There are currently two companies which have been awarded 25-year concession contracts for the West and East zone of the MWSS. Each of these companies is responsible for the operation, the maintenance and the investments in the water supply and sewerage system. The two consortia, the Manila Water Company, Inc. (MWC) and Manylad Water Services, Inc. (MWS) are expected to spend up to US$7 billion during the contract period to improve and expand the system. MWC is a joint venture of Ayala Corp. of the

5 <http://www.wri.org/facts/cs-pdf/philip.pdf>.

6 'MWSS : A Done Deal, A Model in Privatization', Philippines, COURAGE (Confederation for Unity, Recognition and Advancement of Government Employees) (September 1999: 9).

Philippines, United Utilities of the United Kingdom, and the US-based Bechtel Corp. MWS is a consortium composed of the Philippines industrial group Benpres Holdings Corp. and Suez-Lyonnaise des Eaux.

The privatisation of MWSS was based on a plan drawn up by the IFC. IFC's Executive Vice President, Mr Jannik Lindbaek said, 'I am delighted that IFC could play such an important role in this landmark water privatisation, the first in Asia',[7] a declaration which is hardly astonishing given the consulting fees the IFC obtained. In other words, IFC, like the World Bank, functions primarily as an intermediary between governments and public services TNCs. More precisely, IFC was responsible for the following tasks, (1) ensuring that the privatisation satisfied key objectives by carefully structuring the transaction, (2) marketing this privatisation to prospective investors and giving them the opportunity to prepare fully responsive bids, preparing and overseeing a competitive bidding process, (3) negotiating the transaction on terms and conditions most favourable to the Philippines government and (4) guaranteeing fairness and transparency throughout the process.[8]

It is worth analysing the IFC's role in such a privatisation process in a little more detail: as a matter of fact, the IFC, like the World Bank, are not only promoters of privatisation, they are also intermediaries or brokers between a national government on the one hand and a water TNC on the other. This is a process, which may well, above all, serve themselves. Indeed these Bretton Woods institutions do not only make profits from lending monies to a government so that investments in water production and sewerage systems can be made, but they also garner substantial additional benefits from brokering, as seen in the case of MWSS. Not surprisingly, therefore, World Bank records show that it has brokered a total of 2655 transactions (e.g. loans, BOTs, privatisations, etc.) made in ninety-five countries between 1988 and 1993, yielding US$271 billion in revenues for governments. Developing countries accounted for 85 per cent of sales and 35 per cent of all revenue generated, while industrialised countries accounted for $175 billion in sales and 15 per cent of transactions.[9]

Such brokering does not necessarily need to imply consulting fees, as it can also be rewarded by other mechanisms, such as equity. Again, the comparison between MWSS and Aquas Argentina, a private consortium serving the great Buenos Aires area with 12 million people, appears to be particularly revealing, as shown in Table 5.4.

7 'Completion of Manila Water Privatization', IFC Press Release, Washington, DC: International Finance Corporation (1 August 1997).

8 'Completion of Manila Water Privatization', IFC Press Release, Washington, DC: International Finance Corporation (1 August 1997).

9 'MWSS: A Done Deal': 2.

Table 5.4 A comparison between MWSS and Aquas Argentina privatisations

Item	*MWSS*	*Aquas Argentina*
Architect	WB-IFC $6.2 million in consultancy fees	WB-IFC 5% equity
Contractor	– Benpres/Suez Lyonnaise des Eaux (MWC) – Ayala/NWL (Northwest Water + Bechtel) (MWS)	Suez Lyonnaise des Eaux
Length of contract	25 years (renewable for another 25 years)	30 years
Market	11 million	6 million
Profit potential	11 grounds for extra ordinary price increases	$25 million after 20 months of operation
Other incentives	– 8 years tax free – 225% increase in sewerage fee after optional rebasing after 5 years cost in new infrastructure deductible from bonds – Right to government water source for the duration of the contract	
Mass lay-off (%)	57	50+

Source: 'MWSS: A Done Deal, A Model in Privatisation', COURAGE (September 1999: 9).

Institutional arrangements

The current institutional arrangements reflect, in our opinion, the same weaknesses as in the case of Ivory Coast: the strong private actors are not counter-balanced by equally strong regulatory agencies, considering in particular also the fact that administrative structures in developing countries are traditionally weak. The institutional arrangements in the Philippines are as complex as one would expect given the fact that some potentially lucrative water and sanitation systems have been privatised, while the rest has not. This creates in particular, as we shall see, a problem for water resources management at the national level. The institutional arrangement is as follows:

- Two private operators, each holding a twenty-five year concession.
- As a second actor, we should mention the National Water Resources Board (NWRB), which is responsible for the control, supervision, utilisation, exploitation, development and protection of the Philippines' water resources.
- The main regulatory functions in terms of access, quality, and price are determined by a Regulatory Office (RO), whose function is

> to monitor and enforce compliance by the concessionaires of the contractual obligations under the concession agreement, implement rate adjustments, arrange for public dissemination of relevant information, respond to complaints against concessionaires, and prosecute or defend cases before the Appeals Panel. There are approximately 60 employees in the RO, headed by a Chief Regulator and Four Regulators (for technical, financial, and customer service regulations, and for administration and legal matters).
>
> (David, 1999: 91)

- Water-related services outside of MWSS are supervised by the Local Water Utilities Administration (LWUA). This administration has, among other things, the responsibility for funding the construction of shallow wells and deep wells for communal use and low-income households (David, 1999: 85). We thus have in the Philippines, as in many other developing countries where private operators coexist with public services provision, a dual structure of a regulator and an administration, often made even more complex by the fact that water is decentralised. This creates almost unsurmountable problems for the national policy agency – in this case the NWRB – when it comes ensuring sustainable water resources management.

Privatisation and the environment

We shall come back to the possible institutional arrangements for water regulation in Chapter 7. Let us in this section examine, by referring to the case of the Philippines, how water privatisation fares with environmental protection and more generally sustainable development. Many authors and organisations do indeed claim that privatisation is beneficial for the environment. For example, the IFC says: 'privatisation can make substantial benefits for sustainable development' (Hanson and Oliveira, 1997: 34). In the case of the Philippines, it is said that, as a result of privatisation, average water tariffs will decrease by more than 50 per cent from present levels, water pressure and availability will meet international standards within three years, universal water provision will be assured within ten years, and more than 80 per cent of the population will have sewerage and sanitation services within the concession period (Hanson and Oliveira, 1997: 34). Although one might argue that lower tariffs will lead to increased water consumption, and thus unsustainability, it seems nevertheless that these overall objectives, in particular increased sanitation and sewerage services, are ultimately even beneficial for the environment. However, things are not that simple, as shown in the example below.

In July 2000, Greenpeace reported that the La Mesa reservoir, the source of drinking-water for Metro Manilla's 10 million residents, contained toxic substances owing to the run-off of flood waters at the Patayas dump site caused by heavy monsoon rains. Enero Brilliantes, the Greenpeace spokesman, said that the pollutants would eventually poison the waters: 'As of now, water at La Mesa contains metallic substances such as chronium, zinc and alloys 30 percent beyond their safety level. These metals cause cancer to human beings.'[10] As a result, Secretary Alberto Romuladez of the Department of Health called for an immediate investigation, to which Manylad Water Services, Inc. objected, though acknowledging that the water quality from La Mesa has deteriorated because of polluting substances from the dumpsite. However, Manylad also said that they had requested the closure of the dumpsite some time ago. This example clearly shows the complex institutional realities of, for example, promoting environmental protection in the context of privatised companies. This is not to say that the same problem would not have occurred with a public instead of a private company. However, the mechanisms for addressing environmental protection in the context of privatised operators will certainly become more complex.

As a matter of fact, this example shows that simply setting up regulators concerned with monitoring the concession is not enough. In the case of the Philippines, it appears moreover that quality was simply not defined in terms of sanitation targets. As Cristina David in a 1999 report for DANIDA, the Danish International Development Agency, says:

> the exclusion of a segment of the population and the commercial and industrial users of water from the water supply and sewerage sanitation targets leaves a significant part of the population with little government support to ensure good quality drinking water. The concession agreement also overlooks the larger part of wastewater from commercial and industrial establishments, self-supplied water users and those dependent on private works who will have no access to a possibly more efficient and costly means of dealing with wastewater and industrial effluents.
>
> (David, 1999: 106)

On the other hand, one cannot make environmental policy by means of concessions, in particular because such environmental policy needs to be national in scope, while concessions are generally locally attributed. As Cristina David says:

> The public good nature of water and its basic need character, economies of scale and externalities in its production, distribution and consumption, mean that private sector management of urban water resources cannot be

10 Michael Bengwayan, 'Death Dump Polluting Manila's Drinking Water', Environment News Services <wysiwyg://52/http://ens.lycos.com/ens/jul2000/2000L-07-18-02.html>.

> expected to achieve economic efficiency, social equity and environmental sustainability without appropriate contractual arrangements, strong economic and environmental regulations, and other governmental interventions. These concerns are particularly critical with regard to MWSS because of government's weak capacity to enforce environmental regulations in the face of severe water pollution problems and rapid groundwater depletion within the service area, the bureaucratic constraints encountered in developing a strong economic regulatory office, and the large number of urban poor often bypassed in the allocation of water and sewerage services.
>
> (David, 1999: 84)

This therefore again raises the questions of the appropriate level of environmental (and other) policies and regulation, as well as the question of developing comprehensive institutional arrangements. In previous chapters, we have mentioned on several occasions the need for river basin agencies as appropriate water resources management units. Environmental policies appear indeed to be more effective if dealt with at a river basin management level. With regard to privatisation, there is therefore a need to reconcile the private operators (or even the public operators for that matter) with the national regulatory body, as well as with the river basin management authority, if it exists, in order to ensure coherence at all levels. In short, and in the case of the Philippines, we can see that environmental problems call for better regulatory control in terms of provision and quality of services, along with a national or regional policy enforcement mechanism. We shall come back to the institutional mechanisms of ensuring environmental protection alongside private operators in Chapter 7.

South and Central America: the decentralisation challenge

Since the early 1990s, privatisation of water services has been considered by the World Bank as the miracle solution for South and Central America, just as it has been seen as the miracle solution for every country. World Bank specialists say that:

> Despite substantial efforts to improve the quality and coverage of service, one-quarter of the urban population is not connected to a public water system, half lacks public sewerage, and the sewage is virtually untreated. The results are a constant threat to the health of the entire population, a perpetuation of unmet basic needs of the poor, and a steady deterioration of the environment. The failure of the public sector to provide reliable service is largely explained by its inability to operate efficiently and maintain adequately existing water and sanitation systems. In addition,

> the investments for rehabilitation and expansion are very large. It is estimated that the region's countries need to invest an annual $5 billion in water supply and $7 billion in sewerage and sewage treatment over the next ten years.
>
> (Idelovitch and Ringskog, 1997: Foreword)

Furthermore,

> Latin America loses about 9 trillion cubic meters of water each year, about 33 percent of the water collected and treated for public consumption. While it is impossible for water systems to deliver 100 percent of their water to the household tap, Latin America could cut those losses by more than three quarters if it could reach international standards for properly managed and operated water systems.
>
> (Savedoff and Spiller, 1999: 1)

Private sector participation appears therefore as the solution. Simultaneously, as we will see below, decentralisation is also being promoted by the World Bank and implemented in many Latin American countries as part of an overall public sector reform effort.

Not surprisingly, water TNCs are rapidly entering the Latin and Central American markets. Table 5.5, for example, lists Suez Lyonnaise des Eaux's subsidiaries in Latin America.

Despite this evolution, the Inter-American Development Bank and the World Bank consider that private sector participation in the Latin American water sector is not sufficiently rapid, when compared with other infrastructure sectors. This is emphasised by Savedoff and Spiller when they say that

> despite the recent increases in investment in utilities and the rapid surge in private investment in these sectors, investment in the water sector has generally lagged. In the first half of the 1990s, the private sector committed some $35 million to Latin America's electricity sector in 50 projects and $22 million in 85 transportation projects, but only $10 million in 19 potable water and sanitation projects.
>
> (Savedoff and Spiller, 1999: 5)

No doubt, investment and private capital are certainly needed in order to develop water and sanitation for all. However, hardly any studies have yet analysed how such private sector participation and decentralisation can go together. We know that underlying the Dublin Principles is the promotion of a highly decentralised structure, combining local government, community management and private actors. The main idea is that, in order to attain 'integrated water resources management', decisions need to be taken at the lowest appropriate level. Moreover, it is implicitly assumed that centralised

Table 5.5 Suez Lyonnaise des Eaux's subsidiaries in Latin America

Country	*Company*		*% owned*	*Indirect %*	*Via*
Argentina	Aguas Argentinas	Water	25		
Argentina	Aguas Argentinas	Water		6.2	Aguas de Barcelona
Argentina	Aguas Cordobesas	Water		3.7	Aguas de Barcelona
Argentina	Aguas de Santa Fe	Water			
Bolivia	Aquas de Illimani	Water	34		
Brazil	Aguas de Limeira	Water	50		
Chile	Aguas de Valdivia	Water			Aguas de Barcelona
Chile	Aguas Decima	Water	50		
Chile	Aguas Decima	Water		12.0	Aguas de Barcelona
Chile	Aguas Quinta	Water			
Chile	Aguas Quinta	Water		11.8	Aguas de Barcelona
Chile	EMOS	Water Sanitation			Aguas Suez-Lyonnaise
Chile	EMOS	Water			Aguas de Barcelona
Chile	SantiAguas	Water			Aguas de Barcelona
Colombia	Alcantarilla do y Aseo	Water		3.1	Aguas de Barcelona
Colombia	ACUACAR	Water		11.0	Aguas de Barcelona
Colombia	INASSA	Water		4.9	Aguas de Barcelona
Colombia	Lyonnaise (Colombia)	Water			
Colombia	Metroagua	Water		5.3	Aguas de Barcelona
Cuba	Agbar (Cuba)	Water		6.0	Aguas de Barcelona
Cuba	Aguas de Varadero	Water			Aguas de Barcelona
Mexico	Aquas de Cancun	Water		12.0	Aquas de Barcelona
Mexico	Lyonnaise (Mexico)	Water	100		
Mexico	TECSA	Water			
Uruguay	Aquas de la Costa	Water		14.4	Aquas de Barcelona

Source: Table compiled according to Public Services International Research Unit Database (31 August 1999).

governments will not be able any longer to provide water and sanitation services to rural and peri-urban areas. Therefore, as Rondinelli states,

> the high costs and administrative complexities of delivering water in rural and peri-urban areas through central ministries and agencies have led many governments to look for alternatives . . . Self-help programmes, informal sector participation, privatisation of services, user charge and cost-recovery financing, or public–private partnership are some of the ways in which governments in developing countries are experimenting to reduce the costs and increase the efficiency of extending water supplies.
>
> (Rondinelli, 1991: 416)

We are also convinced that decentralisation, along with private sector participation, is the way to go in developing countries, and perhaps even in industrialised ones. However, one must acknowledge that decentralised water resources management can function only if there is a strong central government, which can enforce integrated policies and coherent standards. In many developing countries this is certainly not the case today, which may mean that privatisation and decentralisation combined could well be counter-productive. Indeed, as pointed out by Saade *et al.*,

> although decentralisation processes in the water sector in many countries have been popular since the eighties, successes are few . . . The main consequence of this process from government to community level is that it increases the burden of financial, operational, [and] technical, managerial responsibilities at local level which local authorities and communities do not have the capacity to carry.[11]

One should therefore question, whether decentralisation and privatisation are really complementary. As a matter of fact, the case of El Salvador shows perfectly that privatisation is a substitute for public policy and decentralisation an excuse for political apathy. Consequently, decentralisation, at least in some developing countries, may well end up creating a system of exclusion, where the urban and rural poor will be left on their own, while cities could be served by (expensive) private operators.

Case study: El Salvador

Before we can present the case of El Salvador, we need to briefly explain the concept of decentralisation, first in the water sector in general and then in Central America in particular.

> [D]ecentralisation of water supply and sanitation systems can follow one of the three board models. *Devolution* involves transferring responsibility to local government. This is the model that is most often talked about since it is consistent with municipal decentralization. *Deconcentration* is the transfer of responsibility to autonomous regional offices of a national agency. *Delegation* is assigning responsibility to a third-party organization such as a private company or a regional public company.
>
> (Rosensweig and Perez, 1999: 1)

In Central America, with the exception of Guatemala, most urban areas receive services from national water supply and sanitation companies, although there is some municipal provision of services. Water services therefore rely on a strong centralised structure, shaped by governmental priorities. Indeed, apart from Guatemala where virtually all urban systems are municipally owned and operated, most other countries still have centralised water provision. In Honduras, for example, municipal providers account for 62 per

11 Saade, Blokland and Brikké (2000).

cent of urban water connections, whereas in El Salavador and Nicaragua only 20 per cent, respectively 10 per cent, of the connections are provided by municipalities. The Dominican Republic has no municipal operators at all. Rural water supply and sanitation, however, is generally the responsibility of the rural departments of the state water companies (Rosensweig and Perez, 1999: 5).

Our example will focus on El Salvador and present a study which examines whether decentralising water services does bring about improved municipal management of water resources and higher performance in water delivery. It is also important to mention here that the process of decentralisation of water and sanitation services in El Salvador is part of a more general effort of political and administrative decentralisation, put into place in order to reform public service provision more generally. The motives for such decentralisation of water and other services are numerous, but follow essentially political motives and promises of increased local participation.

History of El Salvador's water provision

In El Salvador, the main body responsible for water resources management was the *Administracion Nacional de Acueductos* (National Water and Sewerage Administration) (ANDA). Although in principle responsible for the water and sanitation services throughout the country, ANDA in practice operates 181 municipal water systems out of the 262 existing municipalities. Of the systems not managed by ANDA, seventy-six municipalities directly operate their own systems, one has established a municipally owned water company, one is managed by a non-governmental organisation (NGO) and three are managed by community-based organisations (Linares and Rosensweig, 1999: 1).

There was therefore in El Salvador, and elsewhere, a debate about whether or not local government was indeed capable of running municipal water services. Some water specialists do indeed think that municipalities do not have the capacity to manage the water network on their own, while others believed that municipal management of water and sanitation services should be linked to efforts to strengthen local government and that local control and accountability are the best way to improve services. The national administration, in this case ANDA, generally also does not want to give up power and control, often preferring private operators to municipal control. Not surprisingly, then, ANDA's position was that the municipalities were not capable of managing a water supply system and that other options, including the use of the private sector, had to be explored. Nevertheless, El Salvador's water decentralisation programme, labelled an Environmental Health Project, started in 1998 with the support of USAID. One must clearly distinguish here between of urban and peri-urban services on the one hand, and rural services on the other.

Decentralisation in urban areas

Quite rapidly into the water and sanitation decentralisation process, it became clear that the municipalities were not interested in the direct administration of services, but in indirect provision through municipally owned companies or through concession contracts to private firms. As a result, the Environmental Health Project identified five potential options for the management of municipal water and sewerage systems:

- ANDA's regional offices
- Municipalities (either direct administration by municipalities, administration by a separate municipal company, or shared administration by the municipality and the municipal company)
- Local organisations such as an association or a co-operative
- Non-profit organisations
- Private for-profit companies under contract to the national government or to municipalities (Linares and Rosensweig, 1999: 10).

It was moreover concluded that the size of the systems was the main factor in determining which options were viable: 51 per cent of the urban centres in the 262 municipalities have populations of less than 2000 inhabitants, and 89 per cent have populations under 10,000; 70 per cent of ANDA's revenue comes from the 300,000 services connections in San Salvador, the capital, and 72 per cent of the labour force also works there. Only thirteen municipalities exceed 20,000 in population, and it was concluded that only in municipalities exceeding 20,000 inhabitants was the private sector option viable (Linares and Rosensweig, 1999: 10). Indeed, such conclusions are not surprising, as private sector participation can be successful only where a profit can ultimately be made. Private operators will make benefits from markets which were previously under government control, whether they have been managed efficiently or not. But this also means that there are fewer options left for the other municipalities, as cross-subsidies from profitable markets to unprofitable areas will no longer occur, if they ever did. Not surprisingly, USAID also concluded that the municipalities not to be privatised will need technical assistance in order to be able to develop the capacity to implement municipally driven solutions. In other words, such decentralisation of municipal water and sewerage systems will require certain investments to start with, which if they do not come from donors will severely handicap the decentralisation programmes.

Decentralisation in rural communities and peri-urban areas

We know that prior to decentralisation, the rural systems were already poorly maintained and developed. As the Environmental Health Project Paper described,

> unfortunately, when the system requires major repairs, many rural communities have only one practical recourse: the municipality. ANDA has not responded in these cases due to lack of resources and capacity. However, most municipalities are already overextended, underfunded, and understaffed and cannot take on this additional burden.
>
> (Linares and Rosensweig, 1999: 6)

The successful transition to a decentralised water system therefore appears difficult. Again, the Environmental Health Project identified three possible options:

- An autonomous organisation (municipal company, private company or NGO)
- A regional office of ANDA
- A national regulatory agency specialising in rural water supply and sanitation that contracts with other organisations such as NGOs or private companies (Linares and Rosensweig, 1999: 10). This third option, according to the study, appears to be the most desirable, even though only NGOs seem to be willing to contract with the national regulatory agency.

Conclusion

Rosensweig and Perez (1999: 6) have studied and compared water and sanitation reform in several Central American countries and come to some interesting conclusions, which we would briefly like to summarise here:

- In all countries except Nicaragua, there is confusion between the functions of *planning*, *regulation*, and provision of *services*. In urban areas, national water companies are responsible for their own planning and are not regulated. Where municipal provision of services exists, the municipal councils decide on investment plans and tariffs and are self-regulating. National planning offices, where they exist, often do not have much to say at the municipal level.
- Effective environmental and other regulations do not seem to exist. Neither the regulatory framework nor its enforcement are protecting water sources, regulating water use, or monitoring the impact of sewerage on receiving bodies. These concerns have only begun to be reflected in the legal framework in Central America. It is not exactly clear which institution will be in charge of developing and providing such regulations.
- Although decentralisation is a stated goal of the state modernisation process, it has not yet translated into a consensus as to how to municipalise responsibility for provision of water supply and sanitation services. In rural and peri-urban areas, NGOs appear to be the most

effective agencies in implementing water supply and sanitation programs.

- As of yet, concrete evidence does not exist that decentralisation improves cost recovery. Municipalities have not shown the inclination to increase tariffs. In fact, they may be more likely to charge below-market rates than state-owned companies would because of local political pressure. In rural areas, community based approach[es] have led to improved cost recovery of operating expenses, but initial capital investments are provided by the national governments and external sources.

(Rosensweig and Perez, 1999: 5–6)

Decentralisation and privatisation, despite the World Bank's affirmations (see Chapter 3), do not necessarily go hand in hand. Rather the opposite may be the case: if decentralisation favours privatisation of the lucrative municipal services and allow for more NGO involved in rural areas, such efforts will immediately call for centralised policies and planning, as well as for a coherent regulatory framework and corresponding national enforcement and controlling mechanisms, as municipalities and even less so communities, will generally be too weak to guarantee policy implementation and services standards. In other words, decentralisation in operations, may therefore well be accompanied by re-centralisation in policy and planning, in standard setting, as well as in controlling and control.

East and Central Europe: privatisation as ideology

Eastern and Central European countries differ from developing countries insofar as their infrastructure is generally much more developed. Indeed, these countries with the help of the former USSR, had developed impressive infrastructure network systems, some of which still are in quite good condition; this is certainly the case in the water sector. Therefore, they do not have the developing countries' problem of access to water and sanitation services. Here, the reasons for private sector participation are much more ideological in nature. Katarina Zajc perfectly reflects this ideological view, which would probably not be formulated differently by a World Bank official:

> to make a leap from centrally planned to market oriented economies and to achieve economic development, economic reforms have to strengthen the continuity and stability of the contractual environment by privatizing enterprises, denationalizing assets in the economy, and liberalizing and deregulating the economy. At the same time, countries also have to reform public administration and try to enact modern regulation in order to guarantee the rule of law.
>
> (Zajc, 1996: 21)

In other words, the reason for private sector participation is primarily a political one, i.e. the transition from communism to market economy.

If one looks at Eastern and Central Europe (CEE), one finds precisely this: an accelerated market transformation scheme pressed upon these countries as a result of ideological victory, rather than as targeted reforms. The water sector is probably among the most perfect illustrations of this imposition, as hydrology has been one of the major specialities of Eastern European countries, especially Russia. Yet, the water sector now finds itself under a considerable pressure to reform. As an illustration, Table 5.6 lists the current World Bank projects in the water sector in Eastern Europe and Central Asia.

Since the early 1990s, one can therefore observe significant private sector participation in CEE countries. The Czech Republic has been one of the pioneers, reforming its water sector entirely during the early 1990s. As a first step, it decentralised, charging the municipalities with the responsibility for water provision and sanitation facilities. This was explicitly done in order to bring the private sector on board. Prague, for example, is in the process of privatising its water services, and, not surprisingly, Suez Lyonnaise des Eaux and Vivendi/Anglian are the potential contenders. Many other projects of private sector participation and joint ventures can indeed be now found in the CEE countries. In Hungary, for example, the city of Szeged formed a joint venture with the Générale des Eaux of Vivendi, and obtained a fifteen-year lease contract. Suez Lyonnaise des Eaux is particularly active in the Eastern European market, especially through joint ventures and participations, as shown in Table 5.7.

In addition to the World Bank and TNCs, one must also mention the European Bank for Reconstruction and Development (EBRD). The EBRD is particularly present through the Balkan Stability Pact. Also, the Municipal and Environmental Infrastructure branch of EBRD, which had invested by

Table 5.6 World Bank projects in Eastern Europe and Central Asia, 2000

Country	*Project*	*Committed amount ($US million)*
Albania	Water Supply Urgent Rehabilitation	10.0
Armenia	PSP Municipal Development	30.0
Azerbaijan	Baku Water Supply	61.0
Bosnia-Herzegovina	Mostar Water Supply and Sanitation	57.0
Croatia	Municipal Environmental Infrastructure	36.3
Hungary	PSP Municipal Wastewater	31.6
Kazakhstan	Atyrau Pilot Water	16.5
Poland	Bielsko-Biala Water	21.5
Romania	PSP Bucharest Water Supply	25.0
Turkmenistan	Water Supply and Sanitation	30.3
Uzbekistan	Rural Water Supply and Sanitation	75.0

Source: <http://www.worldbank.org>.

Table 5.7 Lyonnaise des Eaux's participations and subsidiaries in Eastern Europe, 1999

Country	*Company*	*% owned*
Bulgaria	Safege (Bulgaria)	100.0
Czech Republic	Brno VAK	31.0
Czech Republic	Jihomoravske VaK	34.0
Czech Republic	Lyonnaise (CZ)	100.0
Czech Republic	SMVAK	45.0
Czech Republic	VaK Ostrava	40.0
Czech Republic	Vodarny Karlovy Vary	49.8
Hungary	Budapest Water	12.5
Hungary	Eaux de Kaposvar	35.0
Hungary	Lyonnaise (Hung)	100.0
Hungary	Pecsi Vizmu	48.0
Lithuania	Lyonnaise (Lithuania)	100.0
Poland	Lyonnaise (Poland)	100.0
Romania	Safege (Romania)	100.0
Slovenia	Aquasystems d.o.o	50.0

Source: Table compiled according to Public Services International Research Unit Database (31 August 1999).

the end of 1999 a total of approximately 556 million Euro for financing operations in CEE countries,[12] plays the role of a direct financier for municipalities and private companies in the water and wastewater sectors of CEE countries. A major programme is under way for the Macedonian water sector and major loans for the city of Sofia (Bulgaria) and Tallinn (Estonia) have been approved.

Let us briefly mention here two examples in order to highlight the ideological approach to private sector participation in Eastern Europe. Our first example is Poland, more precisely the city of Gdansk. In July 1992, the city of Gdansk decided to call for private sector participation in the distribution and sewerage of water. As a result of the bidding process, SAUR won the contract and established SAUR Neptun Gdansk. The arrangements were as follows: the city of Gdansk owns 49 per cent and SAUR owns 51 per cent of the equity. The total value of the equity approximates $US 5 million (Zajc, 1996: 114). The type of contract was a thirty-year lease. As such, the city still owns the infrastructure and is responsible for investments, financing, regulation and setting tariffs. On the other hand, it is the newly created company which is responsible for operations and maintenance of the water system, quality of service standards and billing and collection. One can see from this kind of arrangement that the city of Gdansk still carries all the risks, while operational profits will be equally shared between SAUR and the city.

12 'The EBRD: Its role and activities' <http://www.ebrd.com>.

Our second example is Bulgaria. As in Poland, water access is very high, i.e. 98 per cent of the population is connected to a water supply. Since the early 1990s, Bulgaria has, like most other countries, reformed its institutional water management structures. Indeed, an inter-ministerial National Council on Water was established and the Ministry of Regional Development transformed the State engineering, construction and water companies into 'sole proprietorship Limited Companies', and by doing so eliminated operating subsidies. All these moves have been favoured by the World Bank and are aimed at eventual privatisation. The municipality of Sofia has now selected a concessionnaire for its water and sewerage systems. The battle was particularly fierce, which can be explained by the four following reasons:

- First, the state of the water systems requires substantial investments. Therefore, business opportunities are correspondingly high. Indeed, most of the pipes of the water system date back to the 1950s and 1960s, and need immediate replacement. As a matter of fact, Western Europe's water systems also date back to this period (see Chapter 6).
- Second, Sofia is big, accounting for 17 per cent of the country's water market.
- Third, Sofia is unusually solvent compared with other municipalities, and also has a relatively affluent population.
- Fourth, taking over the running of a capital city's water system means stronger credentials when competing for other Eastern European markets.

In this case, EBRD had developed the twenty-five-year concession contract and offered consulting services to the municipality of Sofia. EBRD is also the lead lender in support of investments required in the initial years of the concession, as it will put up $US 35 million of the total project costs of $US 94 million. The borrower will be Sofijska Voda AD, responsible as concessionaire for the provision of water and wastewater services in the city of Sofia. The owners of Sofijska Voda AD are United Utilities plc, Bechtel Entreprises Holding Inc and Edison SpA.[13] To many, this victory over the Franco–German consortium Vivendi/Marubeni/Berliner Wasserbetriebe came as a real surprise.

Again, we can observe the TNCs' strategies in Eastern Europe, which are actually very similar to their strategies in developing countries: not surprisingly, TNCs rely on international financial institutions and some form of local co-operation, in this case on EBRD. One must also note that in CEE countries, privatisation is part of a larger effort of converting former communist countries into full-fledged market economies.

13 EBRD, Sofia water system concession, Bulgaria <http://www.ebrd.com>.

Conclusion

This chapter has shown that, at least in the case of developing countries and transitional economies, private sector participation appears to be an irreversible trend. At the same time, it highlighted the fact that, in most cases, national policies and institutions were not (yet) adapted to incorporating private actors within the overall water resources management framework. What is worse, decentralisation – which is advocated by the World Bank in parallel to privatisation – further exacerbates the discrepancy between powerful TNCs on the one hand and weak municipalities on the other, as shown particularly clearly in the case of El Salvador. But in Africa also, private sector participation appears as a formidable challenge and opportunity – water TNCs will bring, for example, investment flows and know-how, which otherwise can hardly be obtained. And the Manila example shows that the contractual agreement with the water TNC did not include some of the most crucial matters, i.e. environmental protection. Some sort of relationship between the government and the operator proves to be essential, and this is especially true in terms of environmental regulation.

A second problem highlighted in this chapter is the question of how private sector participation fits in with the national structure. It appears that private sector participation is able to develop, at least until now, only in densely populated urban areas, as we have seen in the case of Abidjan, Manila, Gdansk, Prague and Sofia. Only there, it seems, sufficient profits can be made so that the network will not only be maintained, but also efficiently managed. However, where investments are most needed, i.e. in the peri-urban areas inhabited usually by the poor people, the private operator has usually no obligation to provide water services. Integrating urban and peri-urban areas so that they will be served by the same concessionnaire will require solid contractual arrangements, as well as a strong regulator, so that the private operator will provide universal services in terms of access, quality and affordable price. However, it is unlikely that municipalities and cities can set up such strong regulators.

A third problem highlighted in this chapter pertains to the growing discrepancy between rural and urban areas (and more precisely large urban areas), a discrepancy, which is still being enhanced because of the introduction of private sector participation. As shown in the case of El Salvador, rural areas are left to their own fate and not many options are proposed, except for community management or NGOs. This problem should be a main preoccupation, as it hinders the development of the rural areas, and further exacerbates urbanisation. In other words, water privatisation in urban areas cannot be looked at separately from the growing divide between the urban centres and rural areas.

But this divide is precisely what makes it interesting for TNCs to enter the market, a practice one can characterise as 'cream-skimming'. Cream-skimming is the practice of skimming off the most lucrative or potentially most interesting business, while leaving all non-lucrative operations to government, local authorities, communities or NGOs. The fact that the World Bank, and other lenders are also mainly interested in urban areas where return on investment is a likely prospect further exacerbates this trend towards cream-skimming. In the case of CEE countries, one can furthermore detect an ideological pressure on lending agencies and governments, so that lucrative markets be opened up for water and other public services TNCs.

Overall, this chapter has highlighted a serious lack of institutional development and regulatory frameworks, which would back up the process of privatisation. What is worse, simultaneous efforts of decentralisation and public sector reform often further weaken already inadequate public structures.

Chapter 6

Water markets in industrialised countries

> Private sector participation is widely perceived to be the solution to the failure of many publicly owned and managed water utilities to operate efficiently and make the investments required to meet community needs. However, there are no guarantees that privatisation will actually yield the desired performance improvements. Simply converting a public sector monopoly into a private one provides no competitive incentives for the utility to operate efficiently, make appropriate investments or respond to consumer demands. Likewise, privatisation *per se* need do little to improve sector performance if governments are unwilling or unable to tackle such underlying problems as over-manning, uneconomic water pricing policies, financing the provision of public and merit goods, and restricting over-intrusive political intervention.
>
> (Judith Rees, 1998: 4)

Industrialised countries are of course confronted with very different problems than developing countries and transitional economies when it comes to water resources management. In most cases, water and sanitation coverage is not far from 100 per cent and, paradoxically, this is precisely the problem most states and municipalities are now confronted with, given the needs for investment, but also increased financial pressure for operations and maintenance. In addition to having reached a moment where water distribution and sewerage dating back to the 1950s and 1960s now must be renewed, most industrialised countries are also undergoing substantial public sector reforms. State subsidies decline and, as in developing countries, decentralisation policies are being put into place, leaving local municipalities with the burden of infrastructure management and/or development. As in the South, therefore, private sector participation in water resources management appears to be the next logical step.

In this chapter, we will look at both policies and models in the North. In a first section, we will examine the major guidelines and policies promoted by international organisations catering for industrialised countries, namely the Organisation for Economic Cooperation and Development (OECD), the

Council of Europe and the European Union (EU). Though each of these organisations' policies are different, we will nevertheless discover that the Dublin Principles are, as in the case of the World Bank, at the core of the reforms they propose.

Despite these overarching international policies and guidelines, the models used for managing water resources still differ considerably from country to country. In a second section, we will therefore examine selected European models for water resources management, considering that Europe has a long-standing tradition of public services in the water and other sectors. We saw in Chapter 3 that the World Bank preferred what they call the 'French model'. We will thus look at the main aspects of water resources management in France, namely the public–private partnerships (PPPs) developed under the name of 'gestion déléguée' or 'delegated management', as well as the French approach to river basin management. We will also see that, despite its model nature, France has a very weak regulatory structure. Unlike France, the United Kingdom moved in the 1980s towards total privatisation and, as a result of some quite painful experiences, developed an interesting model of regulation, which we will examine in more detail. Finally, we will discuss the German case, which has a very complicated and fragmented structure, involving numerous actors, mainly at the municipal level, i.e. the so-called 'Stadtwerke'. The regulatory system also appears to be complicated and we will ask ourselves whether Germany does in fact constitute a sustainable model or not.

The United States and Canada are about to move towards much more revolutionary approaches and concepts when it comes to water resources management. We will particularly focus on the United States, where big disparities exist between the different states, and where interstate co-operation within the larger federal structure is not always ensured. However, it appears that new forms of 'co-operation' have emerged, leaving us with serious ethical questions, considering that through these co-operations water is now gradually becoming a marketable commodity. We will illustrate this trend in the case of California and in the case of the creation of a trans-boundary US–Canadian water market.

This chapter will therefore not only illustrate how private sector participation is currently changing water resources management in the industrialised countries. By presenting these examples, we want also to highlight the different regulatory structures and mechanisms which are now being applied in the North, and which might, in the future, also find their way to the South. Finally, we will also show that the water TNCs we previously encountered in the South are now expanding to the industrialised water and sewerage markets. This might in fact have been their original idea to begin with, as these Northern markets will almost certainly be much more interesting. But especially in Europe, all these transformations mean a significant shift from the idea of public services provision, and it is therefore interesting to observe how the European Union in particular is trying to reconcile water markets with public services provision.

Towards a standardised approach to water resources management

We have already seen that, in the case of the South, both the World Bank and the water TNCs are furthering a standardised approach to water resources management. The same movement can also be found in the North, where at least three organisations are promoting such a standardised approach following the Dublin Principles. The OECD emphasises the need for real cost water pricing. The Council of Europe, on the other hand, is more focused on local and regional aspects of water resources management and the idea of decentralisation. The third organisation we will discuss here is the European Union. It differs considerably from the other two inasmuch as it has a binding character on the member states. Its role is also different in that it is a technical legislator. As we will see, such technical legislation, especially in the area of environmental standards, has significant impact on private sector participation in Europe.

OECD

The OECD and the Council of Europe are not significant actors when it comes to the practice of water resources management in industrialised countries. However, one should not underestimate their role in policy-setting. Indeed, both organisations seek to bring about cohesion among the different member states, the OECD having a more economic perspective and focusing on water tariffs, while the Council of Europe rather emphasises the social and political perspectives. In some ways, both institutions are representative of the 'Dublin spirit'.

At first, one may think that Europe and North America have overcome the problems facing developing countries. Indeed, the OECD do not really have this tension between environmental and developmental objectives. Although water scarcity is a worldwide phenomenon, the amount of resources available in the North,[1] as well as the more efficient usage of water resources there, in fact lead to fewer environmental pressures and concerns. Also, access to water is much more widespread than in developing countries, as shown in Table 6.1.

However, in the OECD countries, the pressure on the water sector mainly stems from quality concerns, as well as from the necessity to reform state and local structures as a result of economic globalisation. Despite these remarks, one may be surprised to see that it is the environmental unit which is in charge of water issues within the OECD. As in the case of the Dublin Principles,

1 It must be noted that some European countries, in particular the United Kingdom, Germany and Belgium are in a critical situation, as they have an average of less than 2500 m^3 per inhabitant per year. See Barraqué (1995).

Table 6.1 OECD countries' access to water supply and sewerage, 1999

Country	*Water withdrawal % of gross annual availability* (1)	*Access to public water supply* (2)	*Public waste-water plants, % of population connected* (3)	*Country* (4)	*Water withdrawal % of gross annual availability* (5)	*Access to public water supply* (6)	*Public waste-water plants, % of population connected* (7)
Australia	4	...	...	Mexico[5]	15		22
Austria	3	...	75	Netherlands	5	100	97
Belgium	42	98	27	New Zealand	1	...	80
Canada	2	92	78	Norway	...	...	67
Czech Republic	16	86	59	Poland	19	...	47
Denmark	16	90	87	Portugal	12	83	21
Finland	2	87[1]	77	Spain	29	>90	48
France	24	99	77	Sweden	2	87	95
Germany	24	98	89	Switzerland	5	100	94
Greece	13	86	56	Turkey	15	93[3]	12
Hungary	5	97	32	United Kingdom	15[2]	99	86[2]
Iceland	–	...	4	United States	19	...	71
Ireland	2	80	...	G7	12	...	71
Italy	32	98	61	OECD Europe	14	...	62[4]
Japan	21	96	55	EU-15	20	...	73[4]
Korea	29	84	53	OECD Total	11	...	60[4]
Luxembourg	3	99	88				

Notes: . . . Not available.
– Nil or negligible.
1 If the private water supply and sewerage systems for individual houses are included, piped water is available in 97 per cent of households.
2 England and Wales.
3 85 per cent in rural areas and 97 per cent in urban areas.
4 Partial totals.
5 94 per cent of the Mexico City Metropolitan Area (15.1 million residents) is serviced with a water connection either directly to the house, or from a common distribution tap in the neighbourhood.

Source: Cols (1), (3)–(7), OECD, *OECD Environmental Data, Compendium 1999*, Paris: OECD, 1999;
Col. (2), OECD, *Household Water Pricing in OECD Countries*, Document ENV/EPOC/GEEI(98)12/FINAL, Paris: OECD, 1999.

environmental management questions seem to be at the core of OECD water policies. It was in fact not until the late 1980s that water resources management was dealt with within the OECD. In 1987 the OECD published a document on the *Pricing of Water Services* (OECD, 1987). In 1999 it published *The Price of Water: Trends in OECD countries* (OECD, 1999). From these two main documents, one can deduce the approach and input of this organisation into the water sector. This input is mainly economic in nature, pushing for real water pricing. In short, and according to the OECD, the increasing use of water and at the same time its increasing scarcity should be reflected in price.

Not surprisingly, then, the OECD makes the five following recommendations to its member countries:

- To create an *integrated water resources management policy* across the different sectors. Priority measures are to reform tariffs and create appropriate tools in order to measure the real costs of water including, full economic costing which includes environmental externalities, as defined in Chapter 3.
- To launch further *information campaigns* in order to enhance user responsibility and increased environmental consciousness.
- To establish an *institutional framework* defining each actor's responsibility and to create a regulatory action framework.
- To adopt an *integrated water resources management approach*, taking into account in particular producers and the demand side.
- To increase *user participation* in the decision making process.

(OCDE, 1998)

As one can easily see, these recommendations are very close to the Dublin Principles. However, besides these general recommendations, one should also keep in mind that OECD's main target of reform is water pricing, when it says:

> Water pricing is an important element of water management, which in turn is an important element of environmental management. Water pricing is also a key component of the 'water economy' in a given location, and can generate both problems and solutions for various social issues that may arise in the water management process (e.g., poverty). Because of its implication for the achievement of environmental, economic, and social goals, the implementation of appropriate pricing structures and levels for water services is an important element of the search for 'sustainable development'.
>
> (OECD, 1999: 26)

These considerations constitute OECD's main philosophy in water resources management. As a consequence, and from a purely economic point of view, water pricing means basically that water will be allocated to the highest-valued users, i.e. to consider the demand rather than the supply side. As OECD puts it, 'in theory, a pricing system whose charges are equal to the marginal costs of providing the water services will allocate resources more effectively' (OECD, 1999: 27). From a social perspective, water pricing, following this economic logic of reallocation, will imply that the society as a whole will benefit from this economic gains, rather than the main users. From an environmental perspective, water pricing will be beneficial since it includes environmental externalities.

Within this logic, subsidies are not seen as the best way to achieve economic, social and environmental objectives.[2] From OECD's point of view, 'general reductions in water prices shield all consumers (i.e., not just those in need of assistance) from important economic and environmental signals' (OECD, 1999: 160). With regards to privatisation, the position of OECD is also quite clear, when saying that 'the introduction of pricing systems can facilitate the private sector playing a more important role in the development and utilisation of water resources. For example, privatisation can encourage (or deepen) the development of markets for water services, which can in turn make it easier for marginal social costs to be incorporated' (OECD, 1999: 39).

In conclusion, one can therefore observe that OECD has constructed a general policy on water resources management which is quite similar to that of the World Bank. Incidentally, one could also think that the World Bank's policy is just a copy of the OECD model. However, and as we shall see below, there is more diversity among the different OECD countries than the model imposed by the World Bank on developing and Eastern European countries.

The Council of Europe

The role of the Council of Europe is of course quite limited as a political actor. Nevertheless, it is important to mention it, as the Council of Europe has developed some recommendations, which also focus on social and not only on economic aspects. The Council of Europe, through its Parliamentary Assembly, defines general policies and regional guidelines at the European level. Several important initiatives have been launched by the Council in the water sector. As early as in May 1968, i.e. a decade before any other international conference on this subject, the Council of Europe's Committee of Ministers adopted what was called the 'European Water Charter', with its three main ideas – namely, that protecting water is a common duty, that all stakeholders must collaborate in this, and that water knows no frontiers.[3]

2 For further details, see OECD (1997).

3 Address by Mr Jean Briane, Chairman of the Committee on the Environment, Regional Planning and Local Authorities, Parliamentary Assembly of the Council of Europe, Council of Europe representative to the Conference. *International Conference Water and Sustainable Development* (Paris, 19–21 March 1998).

In 1993, the Parliamentary Assembly adopted a so-called 'Freshwater Europe Action Programme' (Recommendation 1224), which emphasised the need for member states:

- To encourage any initiative concerned with water management designed to safeguard its quality and quantity, not only today but also for future generations
- To ensure that the water resources managers – either public or private - exercise a pricing policy determined by the totality of the exploitation costs including, if possible, those induced by sewerage, and fixed to correspond scrupulously to the cost of production according to the basic consumption indispensable for everyone
- To take into consideration the contribution of the non-governmental organisations (NGOs) in the formulation and implementation of the policies regarding protection and freshwater resources management.[4]

In 1998, in collaboration with the International Secretariat for Water,[5] a European Forum on *Water, a source of citizenship, peace and regional development* was organised resulting in the launching of the European Water Solidarity Network and the adoption of the so-called 'Strasbourg Declaration'. The Strasbourg Declaration was structured around five core ideas, namely water democracy, protecting water resources, water services for a fair economy, water as a factor in local and regional planning and water and communication. The most interesting aspect of this declaration is its reference to social concerns

> Water is also a social capital asset whose management must comply with certain criteria of economic efficiency without however turning supply services into a source of speculative profits. The technical and economic viability of water-related infrastructure and employment should be guaranteed. Technical and financial choices must take account of criteria of social and political acceptability. This means (1) striking the right balance between economic responsibility and decision-making powers, by maintaining, as far as possible, the responsibility of local elected representatives and avoiding the privatisation of capital assets; (2) facilitating access to finance for projects which benefit small and medium-sized

4 Recommendation 1224 (1993)1 on the protection and management of freshwater resources in Europe, Parliamentary Assembly, Council of Europe, Text adopted by the Assembly on 1 October 1993 (51st Sitting).

5 The International Secretariat for Water was created at the initiative of OXFAM-QUEBEC following the International Drinking Water and Sanitation Decade. The Montreal charter signed in 1990 establishes the guidelines of this new NGO <http://www.i-s-w.org/charter.htm>.

> co-operative, and mutual and municipal enterprises which create jobs; (3) offering water users the means of comparing traditional techniques with suitable alternatives while respecting the diversity of the management models; and (4) giving an economic value to services that encourage a sparing used to charge for water consumption.[6]

The Council of Europe's position is indeed quite different from those of the other international organisations we have seen so far. Of course, like the OECD, the various declarations of the Council of Europe do reflect the Dublin Principles, especially in the case of Recommendation 1224. But in particular the 1998 Strasbourg Declaration reflects, probably against the background of globalisation, a new concern for social responsibility, local management and diversity. In other words, though the Council of Europe admits the need for adjustment of the water sector, it does call for the preservation of an approach to water resources management which is locally grounded. Unfortunately, the Council of Europe does not have much say beyond its declarations, which moreover are not concretised to the point where one could, for example, see what the Strasbourg Declaration means in actual local management terms.

The European Union

At present, one cannot really talk about a coherent European water policy, which does not mean that the decisions pertaining to water at the EU level will not have profound impacts on the way the European water market will evolve and the different water companies will adjust. As is the case at the global level, the EU was gradually moving into the water sector via environmental concerns and 'legislation'. Indeed, considering that the different member countries, as we shall see below, each have their own often complex and sophisticated approach to water resources management, it was difficult for the EU to come up with an integrated water policy. According to Kaczmarek, the EU might have added some real value to the different member countries if it focused on common problems, such as transboundary river basins or seas (Kaczmarek, 1997). Instead, it has mainly focused so far on issues of water quality.

At first, i.e. during the 1970s, EU legislation focused primarily on the quality of water. Problems regarding water scarcity and re-used water utilisation were basically not addressed. Concern with water significantly increased during the

6 The Strasbourg Declaration adopted at the European Forum, *Water, a source of citizenship, peace and regional development*, organised by the Committee on the Environment, Regional Planning and Local Authorities of the Parliamentary Assembly of the Council of Europe and the International Secretariat for Water (Strasbourg, 12–14 February 1998).

late 1980s and the early 1990s, in parallel to the EU's new preoccupation with environmental issues and problems. Indeed, one can observe that there has been increased legislation on water issues following the 1988 Frankfurt ministerial seminar, with over 30 Directives, most of them pertaining to water quality. As Kaczmarek says, 'the Community's water policy is at best a policy about water quality' (Kaczmarek, 1997 [our translation]). Between 1995 to 1997 the Commission's work in the water sector became a subject of debate and was questioned. For example, the Parliament stressed the need for reform of the European water policy, because the present legislation was *ad hoc* in nature and the EU did not make any efforts to promote a preventive and integrated policy for water resources management. Rather, the EU limited itself to ensuring water quality for certain special users, as well as to preventing some particular pollutants entering the water cycle.

The Commission proposed a new and much more comprehensive Directive on 16 October 1997, which now integrates surface and groundwater management, water use, water quality standards and water pollution control. This Directive also seeks to promote the 'polluter-pays' principle (PPP), as well as the integration of European policy with national and local policies, so that land and water management will take local characteristics into consideration. This Directive, which was drafted by former Environment Commissioner Ritt Bjerregaard, seeks to harmonise water standards across the Union and to protect and enhance the quality and quantity of aquatic systems. However, some controversy has emerged concerning the scope and the legal aspects of its content.[7] If enforced, this Directive will be of central importance for the evolution of water resources management, TNC presence and the survival of many water companies in Europe.[8]

Indeed, the Union should not be underestimated when in comes to the liberalisation of the European water sector. Water specialists like Marteen Blokland[9] consider that no EU Directives will be issued on the liberalisation of the water sector, since many utilities have already been liberalised and the water sector represents minor benefits as compared to the telecommunications sector, for example. We can agree that probably there will not be a Directive specifically on water liberalisation in the near future, even though there has already been a Directive for the whole infrastructure sector that obliges water companies to tender on a regular basis.[10] However, water liberalisation in Europe, we think, will most likely occur via environmental

7 For more details, see 'Lawyers claim water plan is unenforceable', *European Voice* (11–17 May 2000: 5).

8 Proposals for a Water Framework Directive (Commission Proposals COM (97)49, COM(97)614 and COM(98)76; Amended Proposal COM(1999)271).

9 Interview with author (25 October 2000).

10 See Directive 98/4/CE.

legislation. Also, there is no reason why water TNCs should not be interested in the European water market, especially big cities.[11]

Since the emergence of the environment as an autonomous policy area in the EU during the 1970s, legislation seems to have followed a hierarchical and interventionist patterns of policy steering.[12] And as shown by Nigel Haigh,[13] this is particularly the case for water pollution. Indeed, EU environmental legislation defines concrete emission and quality standards and expects the application of best available technologies (BATs). Besides environmental objectives, such legislation also regulates the means by which these objectives should be achieved, therefore reducing the scope for national flexibility. Moreover, such environmental legislation is directly linked to private sector participation – local governments, in most cases, do not have the capacity to ensure the improvements needed in the quality of water as defined by EU legislation. In other words, the Union actively promotes private sector participation through its environmental legislation.

For example, it was estimated that in the case of France a 1991 Directive on the treatment of water needed an investment of 100 billion French Francs by the year 2005.[14] To quote another example, it is estimated that the United Kingdom will have to invest $60 billion in wastewater treatment up to 2010 in order to meet the new European water quality standards. This amounts to about $1000 *per capita*, or about 0.6 per cent of GDP to be spent on wastewater treatment alone over the ten-year period (Serageldin, 1994: 12).[15] In Germany, for example, a World Bank study concluded that the investment targets required to comply with EU standards could not be met through public sector financing alone.[16] It is thus obvious that EU environmental legislation opens up significant new investment opportunities for the private sector,[17] and it is likely that some of this will be the direct result of efficient lobbying by the interested companies.

11 As already suggested in the title of an article by Bernard Cabaret, Director General of Lyonnaise des Eaux-Dumez. Bernard Cabaret, 'Les distributeurs d'eau face au grand marché de 1993', *Courants*, 8 (March–April 1991: 22–8). See more particularly a graph on the evolution of the size of the water market per countries per billion Francs from 1987 to 2000, Pierre-Frédéric Ténière-Buchot, 'Prospective de la gestion de l'eau en France: Un deuxième essai', *Futuribles* (June 1991: 35).

12 Christop Knill and Andrea Lenschow, *New Concepts – Old Problems? The Institutional Constraints for the Effective Implementation of EU Environmental Policy*, paper presented at the Annual International Convention of the International Studies Association (Los Angeles, CA, 14–18 March 2000).

13 Haigh (1996).

14 Kaczmarek (1997).

15 Serageldin (1994: 12).

16 World Bank (1995: 8).

17 See in particular the proposals for a Water Framework Directive (described above) and the new Drinking Water Directive (98/83/EC) which fixes water quality standards as set by the World Health Organisation.

These examples are just another illustration of the profound impact of EU policy on water resources management. Indeed, EU environmental policy accelerates or even forces the transformation of the water sector by introducing a need for private sector participation. If such strong environmental legislation is moreover combined with compulsory tendering, as is likely to happen in the near future, significant new business opportunities will be opened up in the European water market. An interesting World Bank document says that:

> [T]he political economy of this is of course, complex, with many interests served by such a process. The most obvious and visible beneficiaries are the environmental groups, who legitimately argue for an explicit, uniquely high weight to the environment and who are generally satisfied with the results. Less visible and less forthright is the strong lobby of consultants, contractors, operators and professionals who benefit greatly from the enormous sums of money dedicated to the environment'.[18]

In short, one can see how private sector participation in European water is becoming unavoidable, as important investments will be needed to meet the high quality standards set by EU legislation. This observation has also to be put into the perspective of predominant public ownership of water and sewerage utilities throughout Europe: in other words, the need for substantial investments will combine with parallel public sector reform and decentralisation, with the privatisation of water companies throughout Europe as the main and most likely outcome. Figure 6.1 shows the extent to which water supply is currently still publicly owned in Europe, thus highlighting the fact that, with the exception of France and the United Kingdom, all other European public water utilities will most likely come under very serious pressure in the near future.

The European water sector and its different models

Although the Union is trying to harmonise the water sector in Europe by means of environmental standards and probably later on tendering, there is still a notable difference in the way water is managed in the different member states. Indeed, the diversity of the different structures in charge of delivering water and most often also sewerage services are striking. They range from the Dutch *Wateringue*, which basically takes care of everything from wastewater to water recreation, to the German *Stadtwerk*, which not only distributes water but also electricity, gas, cable and others, to the Italian *Azienda munizzipalizata*, a true enterprise (e.g. Barraqué, 1995). In this section, we will

18 World Bank (1995: 11).

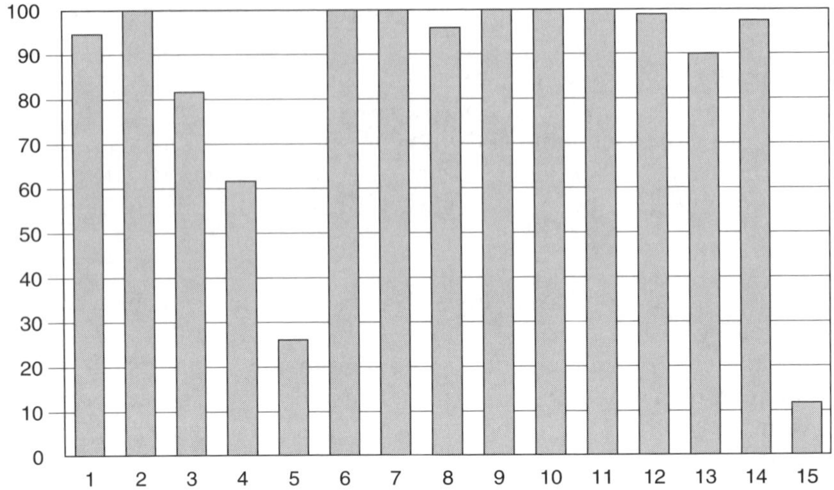

Figure 6.1 Public water supply in the EU (percentage of population supplied by public utilities)

Note:
(1) Belgium; (2) Denmark; (3) Germany; (4) Spain; (5) France; (6) Greece; (7) Ireland; (8) Italy; (9) Luxembourg; (10) the Netherlands; (11) Austria; (12) Portugal; (13) Finland; (14) Sweden; (15) United Kingdom.
Source: Hall (1998b). Published for the International Labour Organization, 1998, p. 122 from Eureau: Management Systems of Drinking Water Production and Distribution Services in the EU Member states, 1996.

present three different models of European water resources management. We think that France, the United Kingdom and Germany represent in some ways most of the different models currently in use. For each of these three case, we will in particular examine the respective contractual arrangements, as well as the corresponding regulatory frameworks.

The French model

France is usually seen as the model for water resources management, in particular when it comes to its river basin management approach and its public–private partnerships in water distribution. As such, the main advantage of the French approach is to combine a well constructed river basin management structure with a strong local decentralised system for the distribution of water. As already shown in Chapter 3, the French model has also been adopted by the World Bank as *the* model developing countries should follow in water resources management.

France has actually a very complex structure with a responsible body at every level. However, it is at the local level where most of the responsibilities are concentrated. The mayor is, as in many other European countries, responsible for the distribution and treatment of water. Mayors have the right to choose any type of contractual arrangement, with a public or a private operator. Mayors are also responsible for investment decisions. Investments are usually made with the financial help of the basin agency, as well as the departmental assemblies. At the departmental level, each department has a *Mission InterServices de l'Eau* (MISE) under the authority of the *préfet*. The Departmental *préfets* carry the responsibility of ensuring that service provision at the community level as respects legal obligations. They can also issue decrees specifying appropriate standards of potable water quality and effluent discharge, in accordance with the national Ministry of the Environment initiatives.[19] This ministry also establishes national guidelines concerning the management of water quality, and monitors the performance of the basin management agencies.[20] At the regional level, it is the *Directions Régionales de l'Environnement* (DIREN) which co-ordinate the activities of the departments. Moreover, it is at this level that 'Funds are allocated on a departmental basis for water and wastewater services. These funds are principally used for monitoring and the provision of free technical and administrative advice to granting authorities at the municipal level'.[21] At the basin level, it is the basin *préfets* who are in charge of implementing the Directives issued by the government's specialised department and who implement the general policies formulated according to the Basin Committee.[22] There are six such river basins in France, so called 'river basin agencies'. The basin and the local levels are certainly the two most important yet complementary levels of this complex system. If the local level focuses on the management and operational aspects, it is at the river basin level that environmental policies are formulated. The basin agencies then implement these policies along the 'polluter-pays' principle (PPP), whereby all water users are obliged to contribute to a kind of environmental savings bank, as illustrated in Figure 6.2.

At the ministerial level, the six river basin agencies are co-ordinated as shown by Figure 6.3.

In addition to this complex institutional structure, France is also known for a successful combination of public–private partnership. Indeed, probably no other country has had such a long experience of private sector involvement in providing water services. As we saw in Chapter 4, such private sector involvement dates back to the beginning of the nineteenth century. The expertise of the French operators when it comes to water distribution is undisputed, and

19 Moraru-de Loë and Mitchell (1993: 137–46).

20 Information based on *La gestion de l'eau en France: une démarche reconnue*, Agences de l'Eau, Ministère de l'Aménagement du Territoire et de l'Environnement (1998).

21 Jones (1997: 27).

22 For more details on the French River Basin Financing Agencies, see Chapter 3 (The Ruhr French Model pp. 88–91).

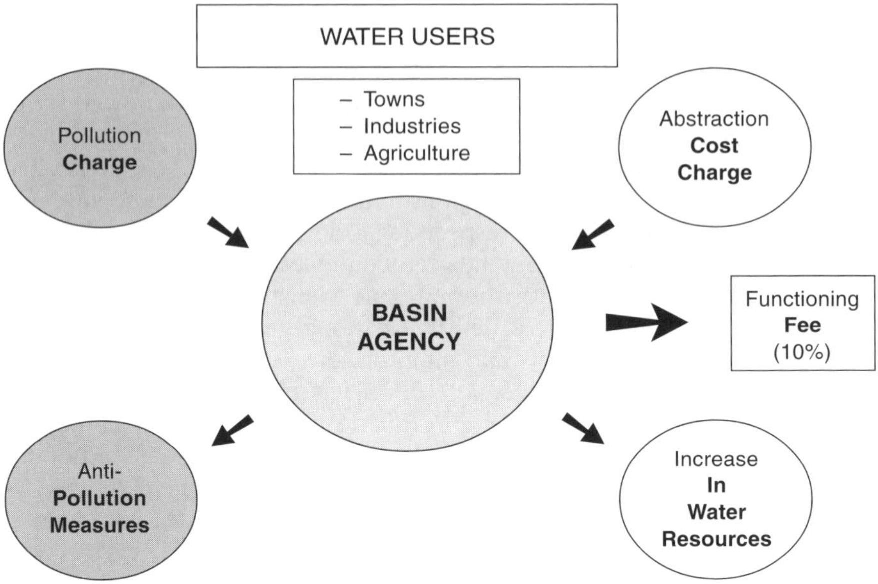

Figure 6.2 The functioning of the French river basin agency.

has led them to become highly efficient. The technological advance of these firms over other competitors is considerable.

The responsibility for the management of France's approximately 12,000 independent water utilities is under the jurisdiction of over 36,000 local municipalities or communes, free, as we have seen, to choose their own contractual arrangements. The power of the French system lies in the fact that there are substantial economies of scale to be achieved, given the high concentration of the privatised water industry. Indeed, today, over 75 per cent of the country's population is provided with water and about 40 per cent with sewage services by private companies, the market for privately provided water being divided primarily among three companies, i.e. Vivendi (35 per cent of the French market), Suez Lyonnaise des Eaux (23 per cent) and SAUR (16.5 per cent).

Not surprisingly, private sector participation in water and sewerage provision all over the world and especially at the World Bank, have been inspired by the contractual arrangements of the French model. Municipalities in France in fact use three general types of contractual arrangements, which vary according to the degree of responsibility assigned to the private companies for managing water-supply systems. These are the concession, '*affermage*' or lease contract, and the management or service contract:

- In a *concession* a private company contracts to finance, build and operate all installations. Bids represent what a firm will charge for water services,

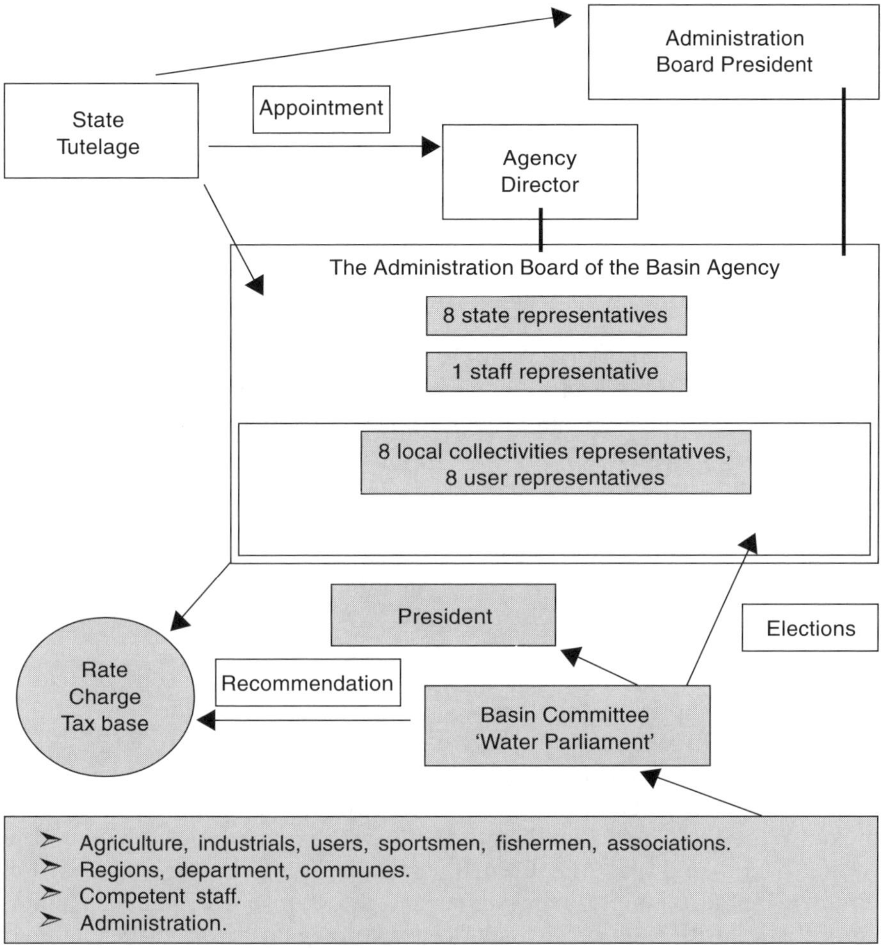

Figure 6.3 The organisation of the French river basin agencies. *Source*: Based on *Organisation de la Gestion de l'Eau en France* (Paris: Office International de l'Eau, 1996).

with the contract duration, which can vary from twenty to fifty years, enabling the company to recover its investments. The private firm is of course also responsible for handling customer relations and billing.

- In the '*affermage*' or lease contract, the municipality finances and builds the facility and contracts out to private companies for operations and maintenance. The contract duration can vary from five to twenty years. Remuneration for the private contractor comes from user fees, which the contractor collects, and reflects full operating costs, plus profit. A 'municipal surcharge' is added to finance fixed assets and is transferred over to the public authorities.

- In a *management or service contract*, the municipality contracts out a specific part of the operations and maintenance services. The municipality retains responsibility for billing customers and remunerating the contractor. The duration of the contract is usually less than ten years.[23]

Overall, therefore, the investments needed are almost equally divided between the water companies and the local communities, with additional funds coming from river authorities, and to a much lesser extent regions and national governments. This means that there are about as many concessions and lease contracts as management contracts in France today.

Despite all these advantages, one must also acknowledge a central deficiency in the institutional features of the French model. Indeed, one may say that the model clearly lacks government control over the utility operators, as there is no national regulator. Everything is left to the communities and the contractual arrangements between them and the water companies. The French model is actually based on two Water Acts of 1964 and 1992, respectively, which define quality and pollution control objectives, as well as adapting national legislation to EU legislation (Serageldin, 1995a: 19). At the national level, the prevailing principle is inter-ministerial co-ordination. Therefore, no single authority responsible for the water sector exists, and each ministry intervenes in the light of its specific competence. In fact, 'regulation' in the French water sector, if it exists at all, relies on the concept of 'global regulation' comprising the three aspects of (1) regulation via political mechanisms, (2) regulation via the protection of professional reputation, and (3) regulation via market mechanisms.[24] This way of 'regulating', of course, overestimates citizens' and consumer control mechanisms and initiatives linked to their satisfaction or dissatisfaction with service provision. Indeed, as stated by Julian Jones, the French system of utility management is thus founded on the assumption that silence indicates consent, i.e. that consumers are satisfied with a service if they do not vote against it at municipal level (Jones, 1997:50). Regulation via market mechanisms cannot be fully efficient either, given the high concentration of the French water industry.

Not surprisingly, then, there have been some distortions in the French water market. As the French Minister in charge of regional development says:

> Average increase in the price of water between 1993 and 1997 has been 8%. This increase has slowed down since, but it is still slightly over 4% per year . . . The price difference between the various management models remains quite considerable. Indeed, in delegated management, the price is generally 20% higher than what it is in direct management.[25]

23 See the different kinds of arrangements in Appendix 6 (pp. 213–47).

24 Lorrain (1994).

25 Compte rendu paru dans *La Nouvelle Républiques des Pyrénées* en date du 14 Avril 1999 <http://www.attac.org/fra/list/doc/foreleau/deputeministre.htm>.

The *Global Water Report* of the *Financial Times* also noted that the prices charged for water services provided by private operators which leased infrastructures tend to be more expensive than those for services provided by municipally operated bodies, adding that the gap had nevertheless narrowed and that private operators were on average 14 per cent dearer in 1997, not 23 per cent dearer as they had been in 1991.[26] Of course, most consumers will not – as had been the case of four inhabitants of Saint Etienne – sue the local water utility by successfully challenging the legality of price increases. But, although the *Cour des Comptes* published in January 1997 a report on the French water industry, criticising the fact that the lack of control and transparency in operations added to excessive pricing, nothing much has been done. The main deficiency of the French model, i.e. the absence of a water regulator, still remains.[27] This weakness, strongly liked by the operators, seems to have not only been copied, but moreover exported to the developing countries.

The British Model

In 1989 the British regional water companies (also called 'water authorities') were sold to private operators. The reasons for this transfer were explained in a government White Paper in February 1986:

- The water authorities will be free of government intervention in day-to-day management and protected from fluctuating political pressures
- The water authorities will be released from the constraints on financing which public ownership imposes
- Access to private capital markets will make it easier for the water authorities to pursue effective investment strategies for cutting costs and improving standards of service
- The financial markets will be able to compare the performance of individual water authorities against each other and against other sectors of the economy; this will provide the financial spur to improved performance
- A system of economic regulation will be designed to ensure that the benefits of greater efficiency are systematically passed on to customers in the form of lower prices and better services than would otherwise have occurred
- Measures will be introduced to provide a clearer strategic framework for the protection of the environment
- Private water authorities will be better able to compete in the provision of various commercial services, notably in consultancy abroad

26 'France/pricing: price increases slow', *Global Water Report*, *Financial Times*, 65 (5 February 1999).

27 According to 'La réforme de la politique de l'eau: Synthèse de la Réforme présentée par Madame Voynet en conseil des ministres' (27 October 1999), a *Haut Conseil du service public de l'eau et de l'assainissement* will be created in order to reinforce transparency and democracy (Ministère de l'Aménagement du Territoire et de l'Aménagement <http://www.environnement.gouv.fr/dossiers/eau/reforme/defaut.htm>.

- Privatised water authorities will be better able to attract high-quality management from other parts of the private sector; there will be the opportunity for wide ownership of shares among employees and local customers
- Most employees will be more closely involved with their business through their ownership of shares, and motivated to ensure its success.[28]

Margaret Thatcher, in discussing why privatisation of the water and sewerage industry was pursued as a governmental policy, explained that

> the privatisation of the water industry was a more politically sensitive issue. Much emotive nonsense was talked along the line of, 'look, she's even privatising the rain which falls from the heavens'. I used to retort that the rain may come from the Almighty but he did not send the pipes, plumbing and engineering to go with it. The Opposition's case was even weaker than this, for about a quarter of the water industry in England and Wales had long been in the private sector. Of more significance was the fact that the water authorities did not just supply water: they also safeguarded the quality of rivers, controlled water pollution, and had important responsibilities for fisheries, conservation, recreation and navigation. It was Nick Ridley – a countryman with a natural feel for environmental issues - who, when he became Environmental Secretary, grasped that what was wrong was that the water authorities combined both regulatory and supply functions. It made no sense that those who were responsible for the treatment and disposal of sewage should also be responsible for regulating pollution. So the bill which Nick introduced also established a new National Rivers Authority. Privatisation also meant that the companies would be able to raise money from capital markets for the investment needed to improve water quality.[29]

As a result, there are ten companies in England and Wales which supply both water and sewerage services. These are the former water authorities. There are also twenty-five companies, which carry water and no sewerage functions. The companies differ in size from Thames Water, which had a turnover of £681 million, an operating profit of £207 million, and serves a population of 7.2 million (1990) to Hartlepool's Water Company, which in 1990 had a turnover of £5 million and an operating profit of £1 million, with a population served of 91,000 (Cowan, 1994: 114–15).

28 'Privatisation of the Water Authorities in England and Wales', White Paper presented to Parliament by the Secretary of State for the Environment, the Secretary of State for Wales and the Minister of Agriculture, Fisheries and Food by Command of Her Majesty (February 1986: 1).

29 Thatcher (1993: 682).

The British model is an interesting one, not so much because of the total privatisation of the water sector, but rather for its regulatory framework. Under the 1989 Water Act, three new regulatory bodies were established to regulate the different aspects of the new private industry. They pertain to (1) the regulation of drinking-water quality by the Drinking-Water Inspectorate (DWI), (2) the regulation of wastewater discharges by the National Rivers Authority (NRA) and (3) the economic regulation by the Office of Water Services (OFWAT). The core business of the private water companies – water supply and sewerage treatment – are licensed by OFWAT for twenty-five years, and in the event of inadequate performance, OFWAT has the power to revoke licenses.

The role of OFWAT is indeed of great interest. Compared to the regulatory situation before privatisation, water licences now follow transparent regulation specified in the licence. The renegotiation of regulations is clearly determined. In addition, water and sewerage utilities now have to release certain data, and the Director General must publish annual reports to satisfy the transparency of the regulation. The details of the regulation of water undertakers are to be found in the licences where, as Cowan puts it, 'charges are controlled by a variant of the RPI-X formula known as RPI+K. There are five services whose charges are subject to the price cap, measured and unmeasured water, measured and unmeasured sewerage, and trade effluent' (Cowan, 1994: 119). Despite intense criticism, especially from the private operators, Table 6.2 seems to show that such a regulatory framework is beneficial, at least when it comes to investment.

Table 6.2 Total investment expenditure in the UK water industry, 1985–91 (£ millions)

Company	*1985–6*	*1986–7*	*1987–8*	*1988–9*	*1989–90*	*1990–1*
Anglian	106.0	123.0	140.7	141.3	192.0	221.6
Northumbrian	39.3	36.9	39.4	41.4	65.9	98.4
North West	163.0	184.3	201.6	216.9	269.4	408.0
Seven Trent	119.8	149.3	159.7	206.9	262.2	394.7
Southern	62.6	65.6	71.2	89.8	125.5	132.6
South West	36.1	37.2	46.7	51.6	80.2	120.0
Thames	108.8	127.1	169.8	184.5	260.8	413.4
Dwr Cymru	54.4	61.7	67.8	89.8	105.3	160.0
Wessex	49.1	59.4	72.8	74.5	100.0	96.8
Yorkshire	94.6	126.7	142.5	148.8	188.2	250.2
Total	833.7	971.2	112.2	1245.5	1649.5	2295.5
Water only Companies	75.4	77.7	88.2	99.8	111.3	196.0
National River Authorities	–	–	–	112.0	131.0	–
Scotland	95.9	104.6	99.5	112.4	128.5	–
Northern Ireland	25.2	25.8	25.3	25.2	27.7	33.0
United Kingdom	1032.2	1179.3	1325.2	1549.9	2048.0	

Source : *Waterfacts*, Water Service Association (London 1991: 44).

The regulator in England and Wales therefore has a very close control of the water industry, not only preventing price discrimination as in the French case, but also monitoring investments. It is clear that the current regulatory framework is positive in that price, environmental and quality standards can easily be controlled and progress towards precise targets can be realised. Nevertheless, it should be emphasised that such a system imposes a certain cost for precise monitoring activities which are paid by the consumers. Moreover, such a system does not really favour competition, a remark which is often made by the private operators.[30]

A German model?

Germany has a long history as a world leader in water resources management. The most famous examples is the Ruhr Basin Water Association (the Ruhrverband). Other examples are the water associations of North Rhine/Westphalia, which have successfully managed small, heavily-industrialised, heavily-populated and heavily-polluted river basins since the 1920s. Today, the German water sector is one of most developed in the world. Germany's waterworks have a lower failure rate in terms of both quantity and quality than any of their counterparts in comparable European countries: Germany's leakage rate is 9 per cent, compared to 19 per cent for England, 25 per cent for France and over 20 per cent for the United States (2000 data) and 85 per cent of Germany's households and businesses are connected to treatment plants. This is the highest rate in Europe and well ahead of England (79 per cent) and France (74 per cent) (Rudolph, 2000b: 2).

Unlike France and the United Kingdom, Germany is a Federal Republic, a fact which significantly affects the institutional structure of water resources management. Indeed, Germany has a decentralised structure with approximately 20,000 autonomous municipalities, running 7000 water supply and 10,000 wastewater utilities (Rudolph, 2000b: 3). These generally take the form of so-called '*Stadtwerke*', which are basically municipal societies operated by the municipalities with independent bookkeeping. According to the German constitution, water supply and sanitation are under the responsibility of the local government. On the other hand, it is the federal government, which is responsible for the legal framework dealing with water resources management through a Federal Water Act, which is transposed to the States' (Länder) through State Water Acts. The federal government also sets the legal framework in the areas of public health and food security (Merkel, 2000: 28–9).

It is therefore local governments in Germany that run the water supply and sanitation services. There are basically two different types of legal arrange-

30 See 'Getting to grips with competition', *Global Water Report*, *Financial Times*, 70 (23 April 1999).

ments of these *Stadtwerke*, which generally also offer other public services, such as electricity, gas, cable, and others:

- Utilities under public law are either part of the local administration, yet with their own budget, or inter-municipal associations co-operating on a regional basis. They are under the control of the Ministry of the Interior, who is the controlling authority of the communes. They sell the water according to statutes or bylaws, charge fees ('Gebühren'), which are not subject to VAT.
- Utilities under private law are organised as joint stock (AG) or limited liability (GmbH) companies, and generally belong to the local authorities. In some rare cases they have other public or at times even private partners (shareholders). These utilities are under the control of the Ministry of the Economy, and regulated by competition law. Their water price is subject to VAT (Merkel, 2000: 28–9).

Among the 7000 local water supply and 10,000 local sewerage companies are 200 utilities with private sector participation in wastewater management and 250 in water supply, ranging from simple management consultancies, to BOT and even the concession model. Almost all of them are very recent, thus indicating a transformation of the German water sector. These 450 utilities with private sector participation serve 40 per cent of the population connected to water supply in Germany. The only example of private sector participation prior to this transformation was Gelsenwasser, a company which operates in North Rhine/Westphalia. Gelsenwasser was founded by a consortium of local municipalities and industrial enterprises in the late nineteenth century. It is a joint-stock company with 22 per cent of its shares publicly owned (by local municipalities) and an additional 28 per cent indirectly publicly owned by public enterprises of the area.[31]

Private sector participation is thus a relatively new phenomenon in Germany: the first Build–Own–Operate–Transfer (BOOT) contract was realised in the town of Algermissen in Lower Saxony only in 1985 (Rudolph, 2000a: 1). The pace quickly accelerated with German unification, and the need for investments and added financial sources. The city of Rostock in the former East Germany is often mentioned as the first example, when in 1992 the city decided to privatise its water and wastewater services in a combination of BOT and concession contracts. Eurawasser – a consortium owned by both Suez Lyonnaise des Eaux and Thyssen, a West German heavy-industry conglomerate – was awarded a twenty-five-year contract for the supply of water and sewerage services in Rostock. This sent shockwaves through

31 World Bank (1995b: 7).

Germany, which suddenly realised that it was not immune from privatisation. The cities of Berlin (Berlinwasser) and Bremen have now followed suit.

There is a question as to whether the German model could be successfully exported in order to acquire a share of the world's privatising water market. It is unlikely that small municipal companies will be able to compete in this global market, and behind Berlinwasser – which, for example, is developing its international business in Budapest and China – one can again find such TNCs as Vivendi and RWE. The most likely evolution in Germany in our view will be the massive entry of the big French public services TNCs, with perhaps a German TNC emerging under the name of RWE, a company which comes originally from municipal energy provision.

It is difficult to talk about a German 'regulatory framework', as the German system is much closer to the French than to the English one. However, it is of course much more decentralised than the French system, given Germany's federal structure, where each Land has its own Ministries playing a role in the formulation of water resources management policies. Nevertheless, although since 1957 a federal law on water resources management (*Wasserhaushaltgesetz*) attributes water policy to the federal government, each Land can still complement and strengthen these federal laws. In addition, three professional associations organised at the federal level also play a key role through defining professional standards. Water quantity in the rural areas is defined and to a certain extent even controlled by the *Deutscher Verband für Wasserwirtschaft und Kulturbau* (DVWK), drinking-water access and quality by the *Deutscher Verein des Gas- und Wasserfaches* (DVGW), and sanitation standards by the Abwassertechnische Vereinigung (ATV) (e.g. Barraqué, 1995: 35). While this is certainly an original model, we do not really think that such professional standards can ultimately replace state sectoral regulation.[32] Also, we wonder whether such 'self-regulation' by professional associations will hold up in a water market dominated by public services TNCs.

The United States and Canada: pointing to the future of water resources management?

Unlike the examples examined so far, the United States is a particular case in that it is a federal state. Not surprisingly, there has always been a debate over water rights between the federal government and the different states.[33] Nowadays, it seems that state law in most cases defines the control, the appropriation and the distribution of water, unless state law contradicts explicit congressional directives. The consequence of that is the retreat of the

32 See for more details 'Regulating for self-evaluation', *Water 21: Magazine of the International Water Association* (October 2000: 36–7).

33 For more details, see Smith (1992: 129–39).

federal state, phasing out grants for the construction of sewerage treatment plants, reducing federal contributions for construction projects, decreasing expenditures for basic water data collection and severely limiting funds for state dam-safety programmes (Smith, 1992: 129).

Another peculiarity of the United States is the disparity between the different states in water resources, which are very unequally distributed. The Southwest, with an expanding population and with the harshest climate, is the area with the most demand for water. The divide is usually the Mississippi River. The lands East of the river receive much more precipitation than lands West of it. Quite logically, Tietenberg points out serious depletion of groundwater in the high plains, California and the Southwest of the United States (Tietenberg, 1996: 199–201).

There seems to be an increasing awareness in the United States that water is scarce and therefore precious. Municipal operators have become a special target of such awareness and concern, as they are considered particularly inefficient. For example, Haarmeyer affirms that

> many municipal water-supply systems in the United States face serious problems associated with capital deterioration, deferred maintenance, unreliable water supply, and underpricing of water services . . . Because municipal water-supply systems, frequently publicly owned and operated, face little capital-market competition and generally cannot go bankrupt, they lack incentives to operate efficiently.
>
> (Haarmeyer, 1992: 1)

As a result, as emphasised by Frederiksen, wherever water is scarce and new supplies are costly or environmentally controversial, the focus becomes 'water-use efficiency', achieved thanks to free market pricing of water (Frederiksen, 1992: 23). Indeed, many lobbies, such as the US Water and Wastewater Summit, the American Water Works Association, or the Reason Foundation promote water privatisation by invoking the lack of efficiency.

A 1987 survey for the US Environmental Protection Agency (EPA) estimates that there are 52,509 community water systems in the United States. More than half of these are privately owned, which includes investor-owned and homeowners' associations, as well as ancillary systems. Small water systems (<3300 population) make up over 83 per cent of all systems. Though privately owned, these water systems or not regulated. On the other hand, most of the water-supply systems serving communities over 3300 people are publicly owned. These generally larger systems serve over 85 per cent of the nation's population (Haarmeyer, 1992: 4). Most lobbies are therefore pushing for the privatisation of these municipal water-supply systems, in particular those of the big cities. Municipal water-supply operators are also pressed to look for additional investments, according to the amendment of the Safe Drinking Act (Haarmeyer, 1992: 7). As with the EU, especially the small operators will be forced to call for private sector participation. Although there are similarities between the United States and the European Union, the US debate is nevertheless significantly different from that in Europe. The following two examples show how water is being valued differently and how the debate follows a certain ideological logic.

California

Our first example is the State of California. California is a water-scarce region and there is a growing need for new high-technology industries. However, historically most of the water has been used for agricultural purposes. As remarked by Dwight Holing, 'more than one out of every two gallons of developed water in the Golden State historically has gone to growing low-value field and grain crops. Producing them requires nearly three times the quantity of water currently used by all urban areas in the state, yet creates fewer than two jobs and less than US$ 250,000 of revenue per every thousand acre-feet of water'.[34] Water reallocation therefore is an important issue, as is the question of water privatisation. Currently, 22 per cent of Californians are being served by investor-owned water companies, as compared to 15 per cent at the federal level (Neal *et al.*, 1996: 1). Lobbies are pushing for more private sector participation. For example, according to a Reason Foundation study, investor-owned water companies in California provide water at the same price as do municipal water companies, even though the former must pay local, state and federal taxes and are expected to earn a profit for their shareholders. The Reason Foundation therefore estimates that cost savings from privatisation range from 10 per cent to 25 per cent (Neal *et al.*, 1996: 8). However, one must note that the main argument against municipal water-supply systems is that they underprice, which means that they cannot finance investments and often not even maintenance costs. This would mean, in fact, that investor-owned water companies fixed even lower water tariffs.

The originality of California's private sector participation in the water sector is the concept of the 'water market'. Water markets rely on a mechanism called a voluntary water transfer, which allows different users to voluntary trade water allotments on a short-term or on a long-term basis, without forfeiting legal entitlement. By matching willing sellers with willing buyers and letting the price adjust to balance supply with demand, a market automatically moves water to where it is most needed. Such market transfers of water can take various forms. As Richard Wahl puts it, they can be short-term (seasonal) leases of water, long-term leases (for example, thirty years), or permanent sales. They may involve transfer of all (or some fraction) of an underlying water right, or they may be transfers of all (or some fraction) of the

34 Holing (1997: 6).

contractual deliveries of water obtained from a wholesale water supplier, such as the California State Water Projects (SWP) operated by the Department of Water resources, or the federal Central Valley Project (CVP) operated by the US Bureau of Reclamation. Transfers may be initiated by owners of private water rights, by water districts, or by individual landowners receiving water within a district. In the latter case, some form of district review or approval is normally required (Wahl, 1993: 4–5).

Environmental protection groups such as the Environmental Defense Fund in principle support the water market. The main reason is that such markets eliminate the need for environmentally costly dams and reservoirs and help to put more water back instream for habitat and wildlife purposes. Up to now, more than 2.3 million acre-feet of water have been exchanged in California, and this practice is now being extended to neighbouring states such as Idaho, Colorado, Nevada and Utah. Along the Rio Grande in Texas, a spot market for water transfer has recently been set up.[35]

Not surprisingly, water trading is increasingly being developed in California, especially since 1991 when the State of California suffered from a serious drought, leading Governor Wilson to address the issue. At that time, a committee was created in order to 'recommend any additional measures which it deemed necessary to combat the drought and protect the health, safety, and property of the people, while providing maximum feasible protection to the important environmental resources of the state' (Wahl, 1993: 9). As a result, the creation of an emergency drought water bank was recommended, which would respond to four critical needs, namely (1) municipal and industrial uses, (2) agricultural uses, (3) the protection of fish and wildlife and (4) carryover storage for 1992.[36] Consequently, the Department of Water Resources was assigned responsibility for operating the water bank. Membership in the bank (as a purchaser) was limited to those who signed contracts to purchase water. Standardised purchaser and seller contracts set out the basic rules for the bank's operation. Water purchases come from three sources, i.e. so-called 'land fallowing' (making surface water supplies available by not planting and irrigating), groundwater (making surface water available by relying on local groundwater sources or by means of direct groundwater pumping), and water already in storage (water reservoirs owned by local districts). Table 6.3 shows how water is allocated to California's drought bank.

The principal groups expressing concern about water banking are above all representatives of agricultural communities worried about the potential for adverse impacts on the local economies (particularly from land fallowing), as well as the impacts of ground water pumping on local water levels and on land subsidence. Representatives of fish and wildlife groups are concerned that

35 See Yoskowitz (1999).
36 See Wahl (1993: 9).

Table 6.3 Sources and allocation for California's Drought Water Bank, 1991 and 1992

	1991		*1992*	
	Acre-feet	*% of Total*	*Acre-feet*	*% of Total*
Source				
Fallowing	414,743	50	0	0
Groundwater	258,590	32	161,593	84
Stored water	147,332	18	31,600	16
Total	**820,665**		**193,193**	
Delta water quality and other corrections	167,137	20	34,478	18
Net supplies	655,528		158,715	
Allocations				
Urban uses	307,373	47	39,000	25
Agricultural uses	82,597	13	95,250	60
State Water Project	265,558	40	0	0
Department of Fish and Game	0	0	24,465	15
Total allocations	**655,528**		**158,715**	

Source: Steve Macauley, Manager, State Water Bank, Sacramento, California (January 1993).

fish and wildlife habitats should be protected when water was transferred. To some extent, these concerns were directed more at the concept of water marketing and prolonged future transactions than at the operation of the 1991 drought water bank. Indeed, where water transfers are short-term in duration (such as transfers under the water bank) or are limited to a small percentage of the acreage in one area, they are unlikely to have severe impacts on third parties even if they involve reduced crop production. The third parties in this case are the business and individuals who supply agricultural producers or who process the output of a given producer. However, if the transfers are long-term and significant acreages of agricultural land are retired, third-party impacts are inevitable. In fact, they are the price society pays for moving water to more economically productive activities.

In short, it seems that market transfers of water definitely have a place in California's water future. Major transfer agreements have become common practice in Southern California. We think that this type of water trading is indeed a revolutionary concept: water is, in this case, no longer a social and an economic good, but becomes a marketable commodity.

Sault Sainte-Marie

This approach could in fact easily expand to other states or even at a regional level between states. This is why the Canadian–US relations with regards to water resources are very interesting. Indeed it is this type of water marketing which Canadian environmentalists fear the most. Canada's water resources are the most important ones in the world, as Canada possesses 9 per cent of the world's renewable water on the planet. And it is clear that pressure on countries like Canada to export its water will only increase in the future. For example, Sergio Marchi, Trade Minister of Canada said in May 1998: 'I feel very strongly about the issue of water. I think it's true that today's water will be tomorrow's oil. I think we should move forward very cautiously.'[37]

Water has already become highly politicised, especially during the debates on the North American Free Trade Agreement (NAFTA), where anti-NAFTA activists deplored the fact that Canada's participation in this trade agreement would limit its control over natural resources, including water. In 1987 the government had proposed a law stating that Canada was opposed to large-scale exports of Canada's water through pipelines, although it did not oppose smaller-scale exports such as those through bottled water. However, the law which was proposed during the federal elections was later rejected. We should also mention that in 1998 alone three cases against Canadian firms or state governments were filed, all concerning water trade rights.

One of these cases was the Nova Group of Sault Sainte-Marie. Nova announced in 1998 that it had received a five-year permit from the Ontario government to draw up to 10 million litres of water a day from Lake Superior. The goal of Nova was to send the water in tankers to Asia. The company has since withdrawn its complaint against Ontario, which in turn had revoked the permit. In December 1998, Sun Belt Water, a Californian company, sued British Columbia before a NAFTA panel for imposing a moratorium on water exports after the company had signed an export contract with a Canadian firm.[38]

The threat on water exports has, however, not disappeared. And although no exports of water resources have been allowed, the idea of such exports has not been yet dismissed. As recalled by Maude Barlow from the Forum on Globalisation, it is essential to prevent the export of water from the great lake basin because, under free trade, 'once we turn the tap on, we cannot turn it off'.[39]

Conclusion

This chapter has highlighted the fact that water resources management is not a problem limited to developing countries. Rather, it is a global phenomenon. Nevertheless, the problems in the North and the South significantly differ: industrialised countries will increasingly be faced with financial problems, as most installations date back to the 1950s and 1960s and need to be replaced,

37 Régnier (2000).
38 Régnier (2000).
39 Canadian Council website <www.canadians.org>.

and as, owing to growing pollution and water use, sewerage and water production costs inevitably grow. In the case of Europe, this will be exacerbated because of stricter environmental norms, and perhaps compulsory bidding. In developing countries, in turn, the main challenge is investments. In both cases, the challenge is above all a financial one. And this is the reason why the private already is and increasingly will be involved in the water sector. However, this debate about water sector reform has gone far beyond financial aspects to question the actual role of government and its 'usefulness' in the water sector and in infrastructure development more generally.

This debate is further fuelled by water and other public services TNCs, which put pressure on governments to privatise. In the case of the developing and Eastern European countries, these TNCs have the support of the World Bank, which has already become their instrument for global water privatisation. In the case of Europe, the corresponding instrument seems to be the EU, which, through environmental technical Directives, has accelerated the pace and scope of water privatisation. The EU has indeed considerably altered the original debate, by emphasising high-quality standards rather than maintenance costs. This can be seen as an odd combination of environmental activists on the one hand and TNCs on the other. Often, both are supported in this endeavour by selected governments such as France, the United Kingdom and Germany which, by doing so, can simultaneously do something for their respective national-based TNCs. In the case of the World Bank, water TNCs also lobby through their governments, but most of the time they have direct access.

This push for private sector participation creates all kind of problems when it comes to the institutional arrangements in the water sector. In particular, it appears that the main reforms have somehow neglected the role of the national level and of the government. Indeed, all the models which we have examined in this chapter lack important institutional features. The French model of public–private partnerships and environmental policy does have very serious problems: we can mention the corruption cases, the collusive links between water TNCs and political parties and the weakness of consumer protection, thus highlighting a clear lack of transparency in the contracts, the pricing and the control mechanisms of these private operators. The United Kingdom which, under ideological pressure, moved to full privatisation, had to learn the hard way. But one can say that, over time, a British model has emerged in which one can now find a quite good balance between the operators on the one hand and the regulator on the other. Nevertheless, we must ask ourselves whether this model will be viable in the long run, as regional private water companies are increasingly taken over by water and other public services TNCs, leading to an increasingly monopolistic market. The German model, as we have seen, is very complex, and shows a clear vacuum when it comes to regulatory structure and corresponding institutional arrangements. And this will become very problematic as German *Stadtwerke* are privatised.

The World Bank, finally, which basically determines the institutional arrangements in the developing and Eastern European countries does not, at the present moment, have a clear conception of how the water TNCs will be regulated, concerned as it has been, so far, in their active promotion. In short, one can see that despite much debate around private sector participation in the water sector little attention has, intentionally or unintentionally, been so far given to question how privatised water is going to be 'governed' in the future.

Considering all these trends, it is therefore essential, in our view, that more attention be given to institutional arrangements when it comes to water resources management. In particular, the questions of regulation and public services provision will have to be tackled, but also the question of the appropriate levels – from local to global – of the corresponding institutional arrangements. This is to what we now wish to turn in our final chapter.

Chapter 7

Re-regulating the water industry

Linking the local and the global

> There is little question that badly regulated private sector participation poses a threat to all users (including poorer households), granting a private firm monopoly rights to service provision. With costs of provision often higher and levels of demand lower in poorer neighbourhoods, private firms are unlikely to have sufficient incentive to improve access in these areas if they are left to their own devices. However, if regulated effectively and if the service provider is innovative, private sector participation need not necessarily be a threat to poorer neighbourhoods.
>
> (Johnstone and Wood, 1999: 2)

In this concluding chapter, we will try to address the consequences of this trend towards privatisation in the water sector. Indeed, as highlighted in Chapter 1, the privatisation of water must be seen as just one element of a much larger transformation of the public sector and the State within the context of economic, cultural and technological globalisations. Consequently, when trying to identify, for example, how the public sector can regulate the privatised water companies, one needs to keep in mind that the public sector is itself struggling to cope with the consequences of globalisation. The goal of this chapter is therefore to develop a coherent conceptual framework, allowing for the regulation of an increasingly globalised water sector in the perspective of sustainable water resources management.

In doing so, we will necessarily question the currently emerging models of institutional arrangements. Indeed, these models, as well as current practices of water resources management, in our view considerably underestimate the dynamics unleashed by privatisation and the subsequent growth of water and more generally public services TNCs. What we are facing today is not just the handing over of water distribution from public to private operators, as the World Bank would like one to believe. Rather, and as part of a much larger process of globalisation, new transnational actors have emerged and grown to the point where they increasingly control such vital goods and needs as water, not to mention the fact that they also control governments. Authors like

Petrella, along with many other anti-globalisation activists, believe that this dynamics can be stopped, and that the status quo constitutes an alternative to water privatisation. We do not agree: not only do we think that the current dynamics in the water sector – and all other infrastructure sectors, for that matter – is irreversible, but moreover we believe that preserving the status quo is an argument that only some rich industrialised countries can afford. Especially for developing countries, change in the water sector is in order, and such change will necessarily involve the private sector. Moreover, the industrialised countries must realise that subsequent private sector participation will not be limited to developing and Eastern European countries. Industrialised countries will be next, which might have the TNCs' long-term strategy to begin with. In the first section of this chapter we therefore want to briefly make our case, namely that private sector participation in water resources management worldwide is an irreversible and unstoppable phenomenon.

Rather than lamenting such privatisation, we think that is now time to pay attention to re-regulating the water sector, so that the worst effects of privatisation – namely, inequities between the rich and the poor, between the urban and the peri-urban and rural areas, as well as the negative environmental consequences – can be stemmed. So far, perhaps with the exception of the United Kingdom's regulatory authorities, most governments, and above all the World Bank, have been mainly interested in promoting private sector participation without worrying much about how to re-regulate the dynamics they have unleashed. This is particularly obvious in the case of the World Bank, which not only actively helps water TNCs to get access to developing and Eastern European countries, but moreover makes sure that these TNCs' investments and interests are and will be protected. In the second section of this chapter we will therefore try to identify the challenges resulting from this omission, deliberate or not.

In the third section, we will try to address these challenges by outlining an institutional framework at the various levels, from the local to the global, whose aim would be to regulate the increasingly globalised water sector with the overall objective of promoting sustainable water resources management. Addressing these challenges purely at the governmental or national level, we think, would be highly unrealistic. Building on the different models presented in this book – e.g. in particular the World Bank, the French, the British and the German models – we will make suggestions as to how to link mainly national regulatory institutions with regional (river basins), local, and global levels and actors. This will also lead us to identifying the main functions such multi-level regulatory institutions have to perform.

Ultimately, however, the development of such regulatory institutions will have to be integrated within a much larger framework of State and public sector transformation, something which we, however, are not able to do within the scope of this book.

Private sector involvement in the water sector: an irreversible trend

Although still highly disputed, public–private sector partnerships are becoming the dominant model not only in the water sector and not only in the developing and Eastern European countries, but in all infrastructure sectors and in the industrialised countries as well. Private sector participation is a very recent phenomenon, dating back essentially to the early 1990s. No matter what indicators one looks at, such private sector participation has only been increasing since then. For example, between 1990 and 1998 the World Bank tracked 1700 projects newly owned or managed by private companies in the water, transport and telecommunications sectors, and the total amounts invested increased from US$16 billion in 1990 to US$120 billion in 1997.[1]

If one looks specifically at the water sector, one finds an even more marked trend. Figure 7.1, for example, indicates the exponential in the amount of public private partnerships in supply worldwide between 1990 and 2000.

And the recourse to private capital is not only limited to developing countries. Figure 7.2 shows that the number of public–private partnerships in the water sector is also particularly significant in Europe and Northern America.

All the trends identified in this book certainly indicate that such partnerships are only in their infancy: Eastern Europe and South Asia are certainly going to constitute important markets for private operators in the near future.

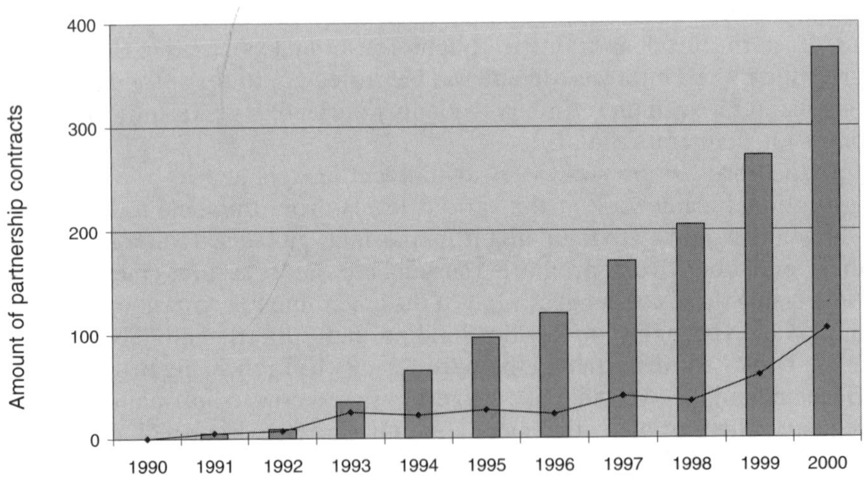

Figure 7.1 Announced water supply by public–private partnerships, 1990–2000.
Source: Franceys (2000).

1 World Bank (2000b: 167–8).

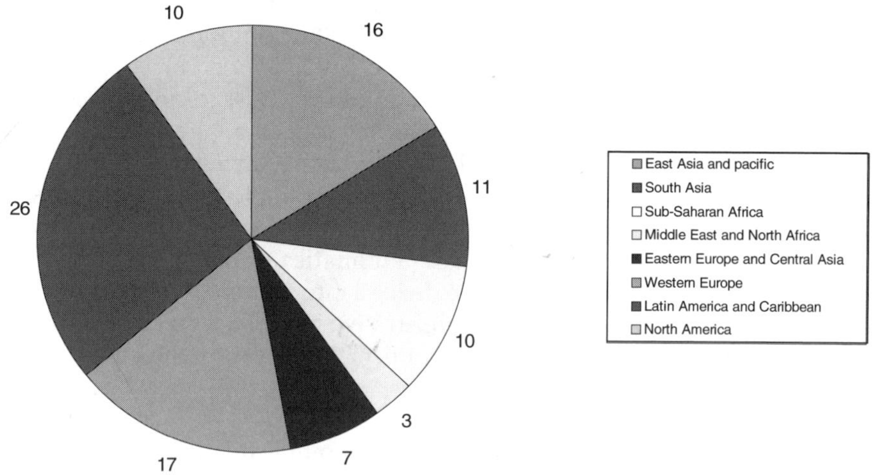

Figure 7.2 Public–private partnerships, clock-wise by region, 2000 (per cent).
Source: Franceys (2000).

Latin America, in turn, appears to be the region which is most advanced when it comes to private sector participation. And, as shown in Chapter 6, the EU's Directive will also bring increased private sector participation. The United States is, without doubt, one of the most attractive markets.

Such facts and figures are proof that the private sector is going to become, whether one likes it or not, a lasting presence in the global water sector. In this respect, Petrella's initiative and arguments, which are quite symptomatic of current opposition to water privatisation, are simply unrealistic. In addition, Petrella also misses the point when he says:

> One has to return the governance of water to its truthful owners, i.e., the inhabitants of the Earth. Water belongs to humanity, it does not belong to the States, the Nation-States. Consequently, it does not belong to the market, nor to the companies, or the shareholders. It belongs to the human community, from the smallest one (the village community) to the biggest one (the global community).
>
> (Petrella, 1999: 127)

We do not in fact think that this argumentation – inspired as it is by New Age principles – is very helpful, as the problem is not a legal one. Nor is it really a question of ownership, as nobody really disputes the fact that the water resources ultimately remain in public hands, and are given to TNCs and other water companies only by concession, lease or management contract. Nevertheless, we can agree with Petrella that such delegated management raises the questions of equity and environmental protection, i.e. social and ecological sustainability.

However, we think that his answer to this concern is misguided, when he says that 'water should be managed by public entities such as local communities, citizens' groups, village or cities networks, or cooperatives' (Petrella, 1999: 29). We have, of course, nothing against co-operatives, even though the water co-operative model has never been applied to large-scale projects. Indeed, one may question whether such a system could work in a megacity, as there exists only one well-known example of co-operative city water management in the world, in the city of Santa Cruz in Bolivia. All of this city's 96,000 domestic customers are automatically members of the co-operative, and the city is furthermore divided into nine water districts, where customers elect members to the administrative board. Several studies at the University of Birmingham[2] do indeed confirm the success of this co-operative management, which would probably be applicable in many small and medium-sized towns.

However, this is not the point. Rather, the point is that Petrella and many others confuse *ownership*, *operations* and (political) *control*. More precisely, they confuse management with control, and think that if peoples are in control of the management they will automatically exert political control. Yet management and politics are two different things. The question therefore is whether peoples – i.e. citizens, consumers, clients, etc. – no matter how well they are organised, should or should not be involved in operational activities. We have of course no argument why they should not, as long as water production and sewerage, for example, are managed efficiently and effectively. However, in light of growing water pollution and corresponding standards, it is not obvious that citizens will automatically be up to the task. In other words, one should not confuse the operation of a water hole in Africa with the management of an urban water system.

In short, the question is not to whom water resources ultimately belong, as this is currently not disputed, even though it may one day become a matter of conflict. Nor is the question really whether or not peoples should remain in charge of political control. Rather, the key question, as we see it, is whether or not the institutional (and corresponding political) regulatory and governance mechanism can be designed in such a way that the private operators, in particular TNCs, can be confined to operation above. Before highlighting what such institutional arrangements would look like, let us briefly crystallise this question more sharply by recalling and summarising our main arguments.

2 See Andrew Nickson, 'Urban water supply sector review, The role of government in adjusting economies', *Paper 7*, Development Administration Group, University of Birmingham; Andrew Nickson, 'Organisational structure and performance in urban water supply : the case of the SAGUAPAC co-operative in Santa Cruz, Bolivia', paper presented at the 3rd International Congress on State Reform and Public Administration (Madrid, 14–17 October 1998).

The main challenges of water resources management today

Throughout this book, we have highlighted the actors and forces which largely explain, we think, the current dynamics of the water sector. Most of these actors and forces are, as we saw in Chapter 1, part of a much larger process of globalisation, which goes far beyond the water sector. The water sector is in fact not even the most 'advanced' sector when it comes to liberalisation and deregulation. In this particular sector's dynamics, it appears, water and more generally public services TNCs, as well as the World Bank and certain national governments, do play a key role. As a result, the way water was traditionally managed is rapidly changing, and urban water supply and sewerage systems in particular are rapidly being privatised. 'Privatisation' here means above all private sector participation in its various forms, i.e. from management contracts, to leases, to concessions. But, as we have argued above, this might just be the beginning of the story: such water privatisation will necessarily lead to profound organisational and institutional adjustments. The first such adjustments will have to occur at the national level, where institutions to manage privatised water will have to be created. But this will rapidly create further institutional dynamics, whereby water issues will also and simultaneously have to be tackled at local and at global levels.

Besides these institutional aspects, one must remember that, at the same time, water resources policies were and still are being reformulated according to a new environmental preoccupation. As explained in Chapter 2, although international meetings had been focusing on water issues since 1977 with the United Nations Mar Del Plata Water Conference, it was only in 1992 that the environmental factor became a main issue within the water agenda with the so-called 'Dublin Principles'. The reason for this rather late awareness is twofold, first the proliferation of conferences in the early 1990s on water as a finite resource. Secondly, environmental disasters, such as the Aral Sea, raised civil society's and NGOs' fears over the way water was being managed. As a result of this newly perceived urgency, a set of policies was formulated during Rio and Dublin, the core of which is the idea that water must be recognised as an economic good. However, neither Dublin nor Rio offered any ideas as to how to deal with the global water crisis from an institutional perspective, let alone a coherent and integrated approach to water resources management. As a first reaction to this 'global water crisis', the international community decided to create a few new organisations, which could hopefully guide the world towards sustainability. Although these organisations (i.e. the Global Water Partnership, the World Water Council and the World Commission on Dams) are still relatively new, one can only be sceptical as to their contribution to sustainable water resources management globally. Their weakness, in our opinion, lies in their absence of institutional mechanisms and in their conceptual confusion. Indeed, they deliberately seek to mix together policy with management considerations. And in the absence of any institutional and intellectual leadership, it is the World Bank, which has filled the vacuum. Today, the globally dominant thinking of how water resources should be managed is indeed formulated by the World Bank, which – at the same time and thanks to its partnership with TNCs – puts its thinking directly into practice.

This is the perspective from which one now has to look at the global water sector today, as there is indeed no other perspective available. Putting this perspective into practice, the World Bank has, over the past forty years, attributed more than $US40 billion of loans to the water sector. This should be proof enough of this organisation's predominance. But the Bank has not stopped at that: reflecting the new division of labour among international organisations, the Bank is now also the world's predominant development agency, thus promoting a corresponding policy. This policy advocates a shift from government-oriented programmes to a much more flexible institutional structure defined around the concept of 'governance'. However, unlike all international organisations, the approach of the World Bank in this regard is much more radical, insofar as the State has been put in question, especially when it comes to State-sponsored economic activities. In particular infrastructure services which were previously under State management owing to their monopolistic nature, must, so the Bank says, be handed over to the private sector. Moreover, according to the World Bank's definition, infrastructure services such as water should be treated like all other goods, i.e. as economic goods. The private sector, the Bank argues, will therefore solve problems the State was not able to solve: not only will the private sector better maintain the infrastructure and expand the network, but moreover will it bring safe drinking-water to another billion people within the next ten years or so.

In this regard, the World Bank's policy is to reform the water sector along two main axes, namely an institutional and legal axis where water users should be empowered to make their own decisions regarding the use of the resource, and a managerial axis where private operators should manage water efficiently. Decentralisation and privatisation thus emerge, according to the World Bank, as the solution to water problems worldwide. While privatisation is generally implemented, the institutional axis generally gets bogged down in some untransparent state reform process. This is not the least, owing to the fact that the institutional French model the World Bank refers to is highly complex and any case not transferable as such to countries, let alone the fact that the model does not even function well in France. As a result, 'decentralisation' most often simply means the opening up of the most lucrative market segments to the private sector, in particular to privileged (French) TNCs.

Private sector participation has thus significantly been facilitated by World Bank policy. What is more, the World Bank has not contributed to empowering local, regional or national water companies. Rather, it has contributed to the emergence and increasingly global predominance of what we have called public services TNCs. Like all other TNCs, the ones active in the water sector display a strong concentration process, with the result that most privatised water supply and sewerage today is under the control of just three or four water TNCs or their subsidiaries or partners. Indeed, one can easily forecast that the water sector will follow the same path as most other globalised industries, where growing concentration is justified by the need to reach a volume critical enough to generate economies of scale. Of course, as we have seen in Chapter 4, these 'water TNCs' have also considerably diversified their activities. Yet, it seems that the tradition of private sector involvement in France has enabled the French water TNCs to acquire their current undisputed role when it comes to water privatisation. At least in the short run, one cannot imagine any serious rivals to Suez Lyonnaise des Eaux or Vivendi. Their multi-service or multi-utilities strategy enables them to offer an integrated package of public services on the occasion of any competitive bid. They can also enter the market in multiple ways, water just being one of the entry points. However, one can observe that these traditional water TNCs are now being challenged by new groups coming from other network sectors, in particular the electricity sector. RWE, thanks to its acquisition of Thames Water, could indeed become the third main global public services TNC, maybe along with American Enron. It is fair to say that it is the World Bank through its policies which has favoured such a concentration process, even though this may not have been its original intention. Though impacting – and, to a certain extent, benefiting – mainly the developing countries so far, this oligopolistic market structure and lack of competition will, sooner or later, also come back to haunt the consumers of the industrialised countries.

Of course, the impact of private sector participation has been, so far, much more important in developing countries and transitional economies than it has in industrialised countries. This can be explained by the combination of World Bank projects and TNCs' interests in the developing countries' megacities. The entry of water TNCs has indeed caused several challenges, the most obvious being the challenge of how to regulate such TNCs in the interest of the consumers (of these TNCs' services) and of the citizens who are left out of the benefits of TNC water provision. This in turn raises the question of corresponding institutional arrangements. Not only have we seen in this book that the World Bank does not have particularly clear ideas when it comes to regulatory institutions, but moreover it is obvious that most developing countries do not have the financial nor the human resources to successfully transform and adapt their public administration and to create regulatory institutions either. Also, the urban poor and as the rural population will not in general get better water supply and sewerage services as a result of privatisation. The opposite might actually be the case, as governments contract substantial World Bank loans to upgrade their urban water systems so that they can be operated by TNCs, money which will not be available to respond to other needs. Such inequity also raises the question of the institutions required to ensure universal service standards in terms of access, quality, and price, at least within a given area, if not across a country. Institutional arrangements generally also remain weak when it comes to environmental and other policies. In both cases, strong enforcement mechanisms are vital when contracting out to private operators. Also, it is not obvious how to apply the new type of contractual arrangements to the local level, given the limited administrative capacity in some countries. Indeed, as recognised by water specialists, successful example of decentralisation are few in developing countries: the main consequence of such a process is that it just shifts the problem encountered by the central government to local municipalities, a problem which is further aggravated by the fact that they have even less financial, operational or technical capacities. Moreover, in the different case studies examined, it appears that there is no concrete evidence that decentralisation improves cost-recovery mechanisms. Indeed, municipalities have no incentive to increase tariffs since they are directly subjected to local political pressure. In short, without strong and innovative institutional arrangements, the entry of the private sector in the water sector appears to be beneficial mainly for urban areas, yet leaves peri-urban and rural areas to their own fate.

This is why we examined in Chapter 6 in more detail the institutional arrangements in these industrialised countries, which already have a certain experience in private sector participation. In fact, private sector participation has developed over time in several European countries, as well as the United States. The diversity of their approaches to water resources management somewhat conflicts with the trend towards standardisation advocated by the OECD or the Council of Europe, both inspired by the Dublin Principles. But the EU is also an active promoter of standardisation. Indeed, through its Directives on water quality standards it directly promotes private sector participation in its member states. More precisely, the Directive[3] seeks to harmonise water standards across the Union, which in the long run will probably end the public monopoly over water supply. In the United States, finally, the pressure to privatise water supply and sanitation must be seen against the background of a general distrust of government in this and other sectors. In light of these trends, it becomes imperative to develop a coherent institutional approach to water resources management, combining private sector participation with a well-defined regulatory framework. The different experiences in European countries have shown that all models still lack important institutional features. Basically, this is owing to the fact that these models have not yet fully taken into account the current dynamics of the water

3 Proposals for a Water Framework Directive (Commission Proposals COM (97)49, COM(97)614 and COM(98)76; Amended Proposal COM(1999)271).

sector with simultaneous tendencies towards localisation (e.g. 'user participation') and globalisation (e.g. TNCs).

To conclude this summary of our book's main argument, one can say that the reform of the sector has not yet been thought through to its logical end, as two important elements are still missing – clarification of the relationship between politics and management on the one hand and a substantial institutional dimension on the other. Indeed, Dublin has proposed and subsequently led to the enhancement of private sector participation in the water sector by allowing water to become considered also as an economic good. But Dublin, and later on the World Bank, have also called for 'decentralisation' and corresponding 'user participation', thus confusing political decentralisation (e.g. user participation) with economic decentralisation (e.g. privatisation). The World Bank, it seems to us, is particularly guilty of this confusion, which is due either to naiveté or to design, or a combination thereof. In any case, neither the World Bank, nor any of these lofty new global water organisations such as the Global Water Partnership or the World Water Council, have managed to come up with a solid idea of how the privatised global water sector can be regulated. Even less have they thought of the institutional arrangements for such a regulation, which, as everybody now concurs, has become a necessity, if not a priority.

Institutional arrangements for water sector regulation today

We now want to outline a proposal for how we think the water sector could or should be regulated so that water resources can be managed sustainably. The particular challenge here stems from the fact that we simultaneously face a trend towards decentralisation, i.e. towards the local level where water supply and sanitation services provision ultimately takes place, and towards globalisation, where the water and other public services TNCs operate. In other words, the water sector stretches from the local to the global, and any institutional arrangement regulating water supply and santitation must take this spread into account. The problem with the World Bank's current approach is that it seems to want to globalise the management of water and to decentralise political control, while we think precisely the opposite should be done. Indeed, the World Bank's often implicit conception is, in our view, a recipe for disaster, given the totally unequal power relationship between local communities on the one hand, and TNCs on the other.

Most of the currently proposed alternative solutions to water resources management also essentially focus on the operator level. Be they economists or political scientists, their solution generally is to replace the existing TNCs by democratic actors, such as, for example, citizens' co-operatives (Petrella, 1999: 29). We certainly have nothing against citizens' co-operatives, but do think that this proposal does not go to the heart of the problem, i.e. the question of the institutional arrangements needed in order to address the *complexity* (e.g. multiple levels) and the *dynamics* of the water sector today (e.g. TNCs). Instead, we propose a conceptual, and corresponding institutional, arrangement which clearly distinguishes between three *dimensions* – e.g. policy and politics, management and markets, and ownership – and four *levels*, namely the local, the regional (river basin), the national and the global levels. Table 7.1 summarises our ideas.

We don't think that the question of *ownership* needs to be discussed here at length. Let us simply stress that one must distinguish between the ownership of the water resources on the one hand and the ownership of the operator on the other. Both have nothing to do with each other and should actually not coincide. While we think that ultimate ownership of the water resources should always remain public – most preferably at the river basin level (which often cuts across several countries) – we don't think that it is likely, and in some cases not even desirable, that the operator is publicly owned. In any case, the regulatory arrangements should be designed so that ownership does not matter.

The core of our argument is therefore to distinguish between four *functions*, each having its own institutional arrangements and financing mechanisms, i.e. a managerial function (local), a river basin management function (regional), a

Table 7.1 Proposed institutional arrangements for water sector regulation

	Policy and politics	*Management and markets*	*Ownership of operator*	*Ownership of water resources*	*Financing mechanisms*
Local	N/A	Private or communitarian	Possible	Possible	User fees
Regional	River basin management involving all stakeholders, including local citizens' associations (French model)	N/A	Unlikely and not desirable	Desirable	Licence fees for using the water resources
National	Sectoral regulation with regional and national consumer participation (British model)	N/A	Possible, but not desirable	Possible	Operating licences and universal services funds
Global	N/A	Competition and anti-trust regulation (US model)	Likely	Excluded	Penalties

Note: N/A = not applicable.

sectoral regulatory function (national) and a competition regulatory function (global). While the first and the fourth functions are more technical and managerial in nature, the second and third functions are eminently political. Only if all four functions are assumed and corresponding institutions established, will there be a possibility for sustainable water resources management. It is also important that all four functions are clearly separated, avoiding in particular institutional overlap and confusion. Let us therefore now present each of these four functions in more detail.

First function: operations (local level)

We have already argued that serious confusion persists between decentralisation and privatisation, leading in turn to a confusion between management or operations on the one hand and citizens' participation and control on the other. This confusion is at times actively entertained by the World Bank and other donors, who want to offer citizens increased participation in exchange for less State financial support. Let us therefore distinguish here very clearly between water supply and sanitation operations on the one hand and citizens' political direction and control on the other. While in operations, especially in small operations, citizens may or may not be involved, they should be involved when it comes to political direction and control. However, we do think that political direction and control should not be directed to the local operator, rather, it should be focused at the regional (i.e. the river basin) and the national levels.

It is at the operations level that one will find the biggest difference between the developing countries on the one hand and the Eastern European and other industrialised countries on the other. Indeed, in developing countries, as we have seen, private operators are basically interested in urban centres (mega-cities), while peri-urban and rural areas would be left to 'user participation'. According to the IHE-WEDC database,[4] among the world's 433 cities larger than 750,000 inhabitants eighty-eight, that is roughly 20 per cent, are currently served by public–private partnerships in water supply and sanitation. On the other hand, among the 40,000 smaller cities and towns in this database, there are only 2350 public private partnerships, representing just 6 per cent.[5] In most cases, at least in developing countries, private sector participation will not be a viable option. In many industrialised countries, and perhaps in some rare case in developing countries, publicly owned local or perhaps regional operators may be able to survive against water TNCs, especially if they themselves diversify into local and regional multi-utilities. However, we anticipate that this will be the exception, rather than the rule.

4 See <http://www.wedc.ac.uk> (July 2000).
5 Franceys (2000).

In many cases these operators will also engage in partnerships with one or the other TNC.

This leaves us with the question of how water supply and sanitation should ideally be operated in peri-urban and rural areas, mostly in developing countries. Significant research has been done here by Tova Maria Solo, a peri-urban water specialist in the World Bank's Water and Sanitation Division.[6] Solo highlights, in particular, the importance of small-scale entrepreneurs in an emerging urban water and sanitation market. These are, for example (Solo, 1999: 121–2):

- individual families with water and/or sanitation connections who provide services for their neighbours
- water points and/or latrines managed or owned by individuals
- bulk water supply systems (tank trucks which distribute to cisterns or to individual families)
- privately owned and managed public toilets and bath houses
- private competitive water networks which provide house connections to water and sanitation
- closed water systems, with treatment plants and piping networks owned by residents
- truckers with private wells
- sewage removal services, septic soil emptiers, night soil carriers, and others
- private waste water treatment plants.

Not much has been written so far on these small-scale entrepreneurs, as they mostly operate outside the formal water supply and sanitation system. However, integrating them more systematically into the water distribution network is quite promising, as shown in the case of Bangladesh reported by Solo:

> In Bangladesh, the water agency has joined forces with private kiosks vendors and with community groups to avoid unaccounted for water. Over the past year, Dhaka has seen the number of privately managed water points in low-income bustee neighbourhoods grow from two to 30. Originally set up through an NGO working under an agreement with the local government engineering department, this programme has left water sales in the hands of either community managed groups or private entrepreneurs who had been accepted by the community. The arrangement favours the water agency which, otherwise, would lose the water to outright theft.
>
> (Solo, 1999: 119)

6 See for example Solo (1999); Solo, 'Keeping Paraguay's *Aguateros* on Stream', *The Wall Street Journal* (Friday 27 November 1998).

Although it appears that there are considerable differences in water pricing, small-scale entrepreneurs nevertheless have considerable advantages over a publicly administered system, in particular the fact that they recover their costs fully and are therefore financially sustainable, that they have virtually no unaccounted-for water, and that they rarely need public subsidies. This, however, does not mean, in our view, that they should not be regulated in some way.

Solo reports another example from Paraguay where 500 small, competing water companies, rely on their own artesian wells and pipe networks, set their prices competitively and finance their own investments. They are very efficient despite their relatively small size and create an impressive competitive environment. Indeed, as stressed by Solo, 'there are few suppliers who do not share several streets with at least one other *aguatero* (private water supplier), and some streets have as many as five competitors' (Solo, 1998). They also prove to be a successful model since they are preferred in most cases to local municipal services. As Solo says:

> In Asunciòn, the private companies also have to compete with the government water company – a difficult rival because, like most municipal water companies, it offers subsidized prices, pays no corporate income tax or water-testing fees, gets an income supplement from government allocations and has the benefits of soft international loans. Yet, many customers chose private suppliers over the cheaper municipally supplied water because, they say, service is more reliable and the water tastes better.
>
> (Solo, 1998)

It is obvious, that from a purely operational perspective, such small-scale entrepreneurs do have significant advantages, as they are much faster than the municipal companies when it comes to providing water to expanding peri-urban areas and shanty towns, as it is especially the case in Latin America.

And these are not isolated examples. It actually appears that such small-scale entrepreneurs today represent a large share of the water markets in developing countries. Indeed, some 25 per cent of the population depends on independent providers for water and about 50 per cent depends on private sanitation businesses in the Latin American cities. In Africa, the numbers are higher with at least half of the families living in the capital cities relying on private suppliers of water and about 90 per cent depending on private sanitation businesses (Solo, 1999: 122). And such a system could probably be developed by means of micro-credits. One can in fact find several examples of micro-financing in sanitation sector,[7] although fewer in the case of water

7 See Darren Saywell, 'Micro-credit for sanitation', *WELL Technical Brief*, Loughborough University (1999).

supply. But if micro-enterprises may constitute a possible solution to the absence of water supply and sanitation operators in developing countries, it will probably be more difficult to the extend the same approach to remote rural areas, where aid may be even more needed. However, we do agree with the Dublin Principle and the World Bank that, even in this situation, one should strive towards some form of user participation, thus considering water as also an economic good, and not just as a free public service.

But, most importantly, one must separate the question of operation – which may or may not, depending on local situations, cover its costs – from the question of political direction and control, even in the case of small-scale entrepreneurs. Wherever possible, some form of competition should be introduced at the operation level, which may be easier with small-scale entrepreneurs than with municipal companies or TNCs. Competition, in our view, is indeed always preferable to monopolies, especially private monopolies.

Political direction and control, we think, should have its roots at the local level, even though, in our opinion, this function should primarily be exerted at the regional and national levels. At the local level, such political direction and control takes the form of citizens' and users' associations. There is a difference between the two, even though, in practice, they will probably most often be identical. If citizens' associations defend political interests and articulate them at the regional level (river basin management; see pp. 225–6), users' associations will articulate consumers' interests and articulate them at the national level. Indeed, citizens and users organised at a local level generally know water and sanitation problems firsthand and are therefore particularly relevant political actors.

However, the term 'user organisation or association' has so far often confused political activism with participatory management. For example, the World Bank's only publication on water and sanitation associations is far from clear when saying:

> Water and Sanitation Associations are broadly defined here as groups of users who act together to plan or provide sustainable water and sanitation services. For sustainable and adequate services to be delivered, the groups have to devise planning, operations, and maintenance arrangements so that the water and sanitation facilities for which they are responsible provide a consistently adequate level of performance over time. To this end, a range of institutional arrangements is appropriate, from informal groups to formal, utility-like organizations.
>
> (Subramanian, 1997: 101)

In the view of the World Bank, these activist water and sanitation associations are in fact service providers, or even service producers. They are seen as another form of operator. However, we think that such associations should have not operational but political functions. They should be seen as lobbies, rather than operators, directed towards both the regional and the national levels. This does of course not preclude the fact that peoples' organisations can operate water supply and sanitation services . . . if they are competitive.

Second function: river basin management (regional level)

The second level required to sustainably manage water resources is the regional level, more precisely the river basin level. These are ecologically and not politically defined entities. It may therefore be the case that river basin management, and corresponding institutional arrangements, cuts across national boundaries, a fact which will pose some serious problems to the political dimension of management. Indeed, even if river basins are an ecological unit, their management will be eminently political, particularly if they cut across national borders. Nevertheless, we are convinced that river basins will become in the future one of the four relevant units of sustainable water resources management. This can actually already be seen in the case of Europe, where the Commission is working towards making river basins relevant managerial units.

The French model of river basin management appears to us as a good start, and should be applied more widely. As we know, the French model foresees so-called 'river basin agencies' in charge of managing water resources within the framework of national environmental and other policies. More precisely, the basin agencies implement these policies on the 'polluter-pays' principle (PPP), whereby all water users are obliged to contribute to a kind of environmental savings bank (e.g. pollution charges and water abstraction costs), from which in turn anti-pollution and more general environmental measures are paid. One could also envisage paying, from this savings bank, services that are in the public interest, such as for example supporting investment in water supply and sanitation systems.

The French have developed a complex institutional arrangement with a politically appointed agency director, an agency board composed of stakeholder representatives (i.e. eight government, eight municipal, and eight user representatives), and a basin committee, which is some sort of 'water parliament'. This institutional arrangement could certainly be improved and adapted to the various river basin specificities around the world. However, we do think that the idea of a river basin agency of sorts is a useful one, and that it is at this level that the so-called 'services in the public interest' ('services d'intérêt général'), such as environmental protection, water supply and sanitation development, and even projects[8] must be defined, managed and financed. Such financing could occur, as in the French case, through pollution

8 One could indeed even imagine that the river basin agency would leverage monies through partnerships, and by doing so engage in physical projects.

and water abstraction costs, but could ideally include also the price of the water concessions. This, however, would mean, that the ownership of water resources would have to be transferred (from national and local public authorities) to the river basin agencies.

Third function: sectoral regulation (national level)

If the World Bank has accepted the French model of river basin agencies – which it sometimes calls Autonomous River Authorities (ARAs) – it is, however, much more reluctant to accept the British and more generally the EU's model of sectoral regulators, which we however think are absolutely necessary for a coherent institutional framework of water resources management. Let us therefore first briefly present the Bank's view on regulation, to which we will oppose the European view, and then add the British specificities.

World Bank

For the World Bank regulation must be seen in the context of development, where development is equated with promoting private ownership and investment by the private sector, generally Foreign Direct Investment (FDI) in the infrastructure sectors. Consequently, the purpose of any regulation is to support successful economic development. This means in particular that investors should be given sufficient guarantees and protection for their investments to ultimately bear fruit (both for the investors and for the country). Furthermore, regulation is seen, by the World Bank, within the larger framework of government intervention in the market. More precisely, it is seen as a significant step in the reduction of government intervention in the development process: from running its public enterprises (i.e. total intervention), national governments now must reduce their interference, set up regulators, and offer the necessary autonomy for private investors, so that they can simultaneously make a reasonable profit and contribute to a country's development. The basic tradeoff a regulator has to face is between short-term private profits and long-term collective benefits.

Implicitly, this perspective says that the absence of development is due to excessive or the wrong kind of government intervention. Therefore, setting up regulators is a means by which government reduces its intervention and lets market forces operate more autonomously. Consequently, in the view of the World Bank, regulators must be agencies independent of government in order to shield investors from (arbitrary, inexpert, or corrupt) government intervention. Independence is moreover justified by the fact that their role is above all as a technical one, as they have to determine the (technically) optimal balance between private investors' incentives on the one hand and national development objectives on the other. This optimal balance can best be determined by economic calculus.

Finally, the World Bank expresses a preference for multi-sectoral competition regulators, i.e. regulators which cover several infrastructure sectors simultaneously and combine sectoral and competition regulation. Again, this can be explained within the World Bank's conceptual framework: multi-sectoral regulators are likely to be less interventionist than sectoral regulators, and they are also likely to be bigger agencies, and thus more protected from government interference. Two aspects of the World Bank's view on regulation thus need to be particularly highlighted. First, regulation is seen as a significant step in the process of removing government from excessive and arbitrary interference into national development. Secondly, regulation is not primarily seen as mechanism of guaranteeing the provision of public (infrastructure) services. Rather, it is seen as a contribution to a development process which, in the long term, may contribute precisely to providing such public services. In other words, for the World Bank, the main function of regulation is 'investor protection'.

European Commission

The European Commission's (EC) chief objective is economic integration. Such integration is considered to be more easily achievable through liberalisation than through standardisation. These efforts have been launched by the Commission, especially since 1986 (Acte Unique), in order to secure the free movement of goods, services, capital and persons throughout the territory of the Union. In particular the liberalisation of services is of concern to the EU, such as electricity, transport, telecommunications and postal, and perhaps water services in the future. However, the Commission considers that the utilities sectors has two specific elements with concrete implications for regulation: first, and owing to the infrastructural nature of these sectors, a natural monopoly will most likely continue to exist beyond liberalisation, leading to imperfect competition, and thus to the need for regulation (Martinand, 1995; Bauby, 1998; Lévêque, 1998). Secondly, and owing to the history of the European social state, any form of regulation will have to pay particular attention to what is called 'public service' or, depending on the sector, 'universal service'. In some European countries public service is a particularly sensitive issue, hence the need for a pragmatic and stepwise approach to the liberalisation of the infrastructure sectors and network industries, respecting in particular national specificities. Moreover, given the EC's pragmatic stepwise approach, the *regulator* is, at least for the time being, a national one with two clear tasks, (1) to promote and guarantee a fair *competition* between the operators, and (2) to control the accomplishment of the *public service missions*, which differ from sector to sector. The European Court of Justice (ECJ), in turn, guarantees Europewide compliance with the various Directives pushing for liberalisation in the different sectors. In short, the Commission suggests a sectoral and independent regulator at the national level, complemented by some action at the European level. A *sectoral approach* is required because of the sector-specific public service requirements, while the need for an *independent regulator* stems primarily from the fact that State-owned enterprises continue to exist alongside private and/or foreign operators. Indeed, the Commission has made it clear that there is no obligation for national governments to privatise their public utilities, which is confirmed by the fact that all (sectoral) liberalisation Directives deliberately avoid the question of ownership.

Three aspects thus need to be highlighted concerning regulation in the EU:

- First of all, the introduction of regulation is seen as part and parcel of a process of *economic integration and harmonisation*, in this case of 'services'. Since each European country has a particular history in providing public services, such integration is a highly political question. Consequently, any regulation – even if the regulator remains a national one – has to address the issue of public or universal service provision, which differs not only by country but also by sector. Sector-specific public service provision is a particular challenge for regulatory design, especially in light of the articulation between the sectoral and the competition regulators. And this is a significant difference from the perspective adopted by the World Bank. Indeed, for the World Bank, sectoral regulation is in essence competition regulation, while for the Commission sectoral and competition regulation pursue two different purposes.
- Secondly, the need for regulation in the case of the Union also stems from the fact that private operators, as well as publicly owned operators from other countries, increasingly exist alongside historical and often still public operators. Rather than the State withdrawing from interfering into the market (which is the view of the World Bank), the Commission considers that the State has to *qualitatively change its way of operating*, i.e. by putting itself at a distance from the various operators, one or more of which are often owned by the same State or other public authorities. The regulator, in the case of the Commission, is seen as creating precisely this distance.
- Both of these considerations thirdly have significant consequences for the way regulatory institutions are designed. Indeed, in the case of the Commission, the regulator needs to be independent of politics (or rather the owner) for a different reason than is the case for the World Bank: since publicly owned operators continue to exist alongside private ones, as well as alongside public operators from other countries, the regulator, even if it is only a sectoral one, has to ensure *fair competition* among the operators. Regulation, in the case of the Union, is thus mainly done in the interest of the consumers (who want the cheapest price), while regulation in the case of the World Bank is mainly to protect private investors from

arbitrary political interference. In short, if EC regulation is consumer-based, World Bank regulation is producer-based.

Britain

The British model adds still another component to this European approach to sectoral regulation, i.e. consumer representation. Indeed, the British water regulator – OFWAT – created in March 1996 a so-called OFWAT National Customer Council (NCC), composed of the chairmen of the ten regional Customer Service Committees (CSCs), plus OFWAT's Director General. The purposes of this Council are manifold and typically comprise the following:

- to represent the interests of all customers, business and household, including those who have special needs
- to assist the Director General in complying with his statutory duty to act in the manner he considers best calculated to protect the interests of consumers
- to be consulted by and/or make representations to the Director General on any matter affecting the interests of customers
- to stimulate policy development by OFWAT and support policy implementation aimed at strengthening the protection of customers' interests
- to assist the Director General in monitoring the performance and outputs of the companies in carrying out the functions permitted in the setting of new price limits, in particular with the Director General's support by means of CSC audits of the companies.[9]

Since its creation, the NCC has been very active,[10] and has successfully defended customer interests in water resources management. It has also played an important role in matters such as quality standards and customer complaints. Giving consumers and their organisations a voice, as well as the right to be informed, can and does make water utilities more responsive to customers, reduces the risk of regulatory capture, increases accountability of regulatory authorities and ultimately makes the regulatory process more effective.

Confronted with the EU's approach to sectoral regulation and to the UK's institutional arrangements in matters of customer protection, one can see that both the French model and the World Bank's approach are grossly

9 OFWAT, *Ofwat National Customer Council Annual Report 1994–95. The Voice of Water Consumers* (Birmingham: Ofwat, July 1995).

10 See OFWAT *National Customer Council Annual Reports* (1993–2000) <Ofwat national customer council website>.

inadequate, not only in industrialised, but also – and perhaps above all – in developing countries. If the French model does not foresee a sectoral regulator and World Bank confuses sectors and sectoral regulation with competition regulation, the institutional arrangement we foresee at the national level should be as follows:

- an independent sectoral water regulator at the national level, backed up by strong and institutionalised consumer representation
- sectoral regulation seen as a political mandate (grounded in a corresponding law), and not as a technical (economic) activity; as such sectoral regulation should guarantee a politically defined standard of public services provision or universal coverage
- such universal service must cover the three following dimensions, namely access to water and sanitation, water and distribution quality, and affordable price.

Throughout this book, we have argued that private sector participation and increased competition are necessary factors for improving water supply and sanitation system networks in developing countries, and to a lesser extent also in industrialised countries. Private sector participation should thus not be put in question and is, as we have seen, in any case unavoidable. However, we are, like many other specialists, in favour of a strong national regulatory framework whose main purpose is to protect the interests of consumers, including the interests of the poorest peoples. This may go as far as to oblige operators to provide certain basic universal services for free, which can also be called 'safety net regulation'. Table 7.2 summarises the main functions or tasks of such a national sectoral water regulator and ranks them according to the type of public–private partnership, i.e. from an operations and management contract to full privatisation (divestment).

Table 7.2 and the preceding arguments show that sectoral regulation must be an important part of any comprehensive institutional arrangement when it comes to ensuring ecologically, economically and socially sustainable water resources management. Needless to say, especially in developing countries and transitional economies – i.e. in these countries where the World Bank is most active – we are still very far from such a sectoral regulator.

Fourth function: competition regulation (global level)

Most countries, today, do have some sort of competition regulator, the strongest one, and so to speak the model, being the US Anti-Trust regulator. In most industrialised countries, this competition regulator is also separate from the sectoral regulators, where they exist. However, and especially in the light of Chapter 4 of this book, we must question whether national anti-trust bodies can be sufficiently strong to regulate such TNCs as Suez Lyonnaise des

Table 7.2 Regulation and customer protection

Regulatory tasks	*O & M*	*Lease*	*Concession*	*BOT*	*Divestment*
Price control	–	✓	✓	✓	✓
Promotion of operating efficiency	–	✓	✓	✓	✓
Service standard specification and monitoring	✓	✓	✓	✓	✓
Control of externalities	✓	✓	✓	✓	✓
Maintenance of public goods functions	✓	✓	✓	✓	✓
Ensure asset serviceability over time	–	✓	✓	✓	✓
Ensure development of essential infrastructure	–	–	✓	–	✓
Controls over powers to manipulate land values/land speculation	–	–	✓	–	✓
Controls over unfair trading practices	✓	✓	✓	✓	✓
Safety net regulations	✓	✓	✓	✓	✓
Promote water use efficiency	✓	✓	✓	*	✓
Ensure responsiveness to final customer needs	✓	✓	✓	–	✓

Notes: O&M = Operations and Maintenance; * = Possibly.
Source: Rees (1998: 22).

Eaux, Vivendi, and others, and whether an international (i.e. globally operating) competition regulator would not be in order.

Indeed, in the water and probably also in other sectors, there is, we think, a growing need for such globally applicable competition and anti-trust regulation. Such a regulator would have mainly two functions, (1) to supervise and prevent collusive practices among TNCs, in our case especially during the bidding process, and (2) to check excessive concentration in both water supply and sanitation, but probably also in public services provision more generally.

Indeed, one of the main problems of recent economic globalisation is precisely the fact that many TNCs are increasingly becoming more powerful than nation-states. While we do not believe that such economic globalisation calls for a 'world government', we nevertheless think that the function of regulating competition and concentration must be assumed at an international level if one wants it to be effective. At present, such global competition regulation seems more or less illusary, even though we can think of several international organisations which would love to assume this function. Let us mention, for example, the World Trade Organisation (WTO), as well as the International Organisation for Standardisation (ISO). For various reasons, we do not think that this is a feasible solution, nor is it desirable. One can also imagine two competing international competition regulators, one being the United States and the other the EU, which is probably a much likelier scenario.

In any case, the function of competition regulation is as essential to a comprehensive institutional framework for sustainable water resources management as are the other three functions (e.g. operations, river basin management and sectoral regulation). Moreover, yet unlike the other three functions, this function of competition and anti-trust regulation must be globally applicable if one wants it to be effective. For the corporations, which are increasingly operating in the water supply and sanitation business, are global operators who have far more power than any other actor within this framework which, in lieu of conclusion, is graphically summarised in Figure 7.3.

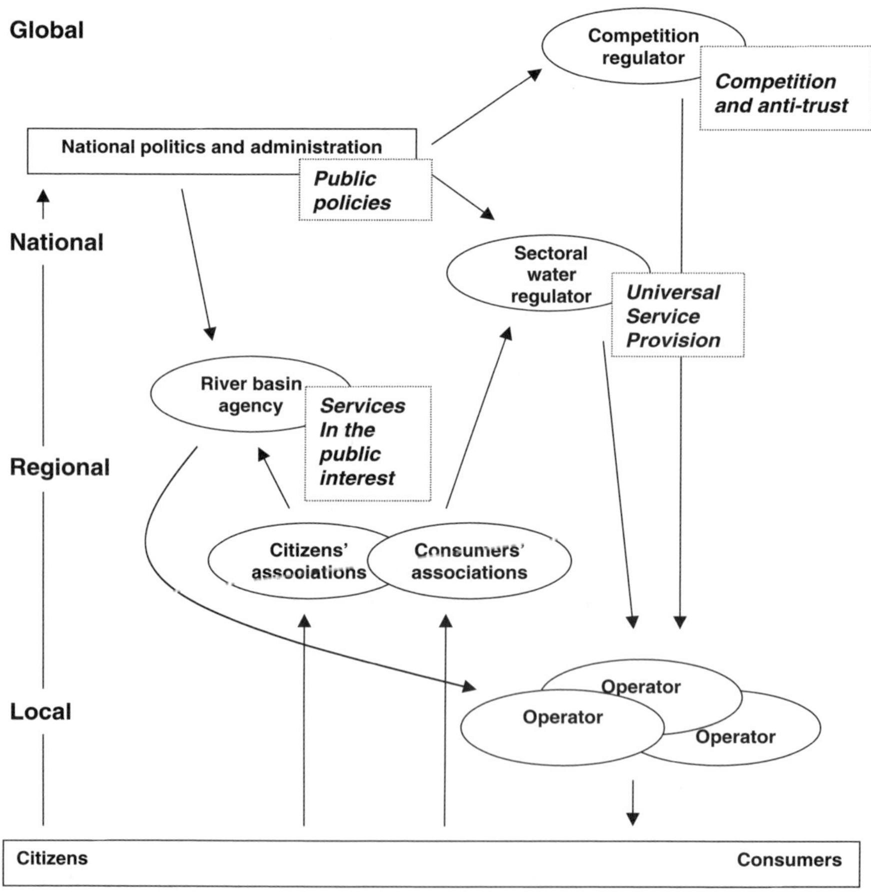

Figure 7.3 A comprehensive institutional framework for sustainable water resources management

However, let us be clear that such an institutional framework, as comprehensive as it is, is not in itself a guarantee that water resources will ultimately be managed in more sustainable way. Rather, it constitutes a system of 'balance and power', whereby the interests of the major stakeholders (citizens, customers, the environment, the operators) are represented and articulated in a way that can somehow accommodate them all. In this sense, this institutional framework could also be seen as a comprehensive water governance system, moreover a governance system that is dynamic and adaptable to the changing needs of society and peoples. If industrial society and the peoples that compose it ultimately want to abuse water resources, this institutional framework will, at best, slow down this evolution . . . but it will not stop the process by itself.

Appendix 1

Steering Committee of the GWP, 2000

Members:

Abdel-Magid, Isam Mohammed, Sudan University of Science & Technology, Sudan
Catley-Carlson, Margaret, Chairperson
Dillon-Ridgley, Dianne, UN Headquarters Representative, USA
Diwan, P.L. Water and Power Consultancy Services, India
Flor, Mai, Lyonnaise des Eaux, Philippines
Forde, Lester, Water for the People, West Indies
Garrido, Raymund José Santos, Dept. Gestao de Aguas Federais de Meio Ambiente, Secretaria de Recursos Hidricos (SRH-MMA), Brazil
Khan, Shoaib Sultan, National Rural Support Program (NRSP), Pakistan
Khatib, Hisham, World Energy Council, Jordan
Van Koppen, Barabara, International Water Management Institute (IWMI), Sri Lanka
Lum, Ken, Commonwealth Science Council, Commonwealth Secretariat, UK
Schreiner, Barbara, Department of Water Affairs and Forestry, South Africa
Zupan, Martina, Hydrometeorological Institute of Slovenia, Slovenia

Co-sponsors:

Briscoe, John, The World Bank, USA
Hodges, John, Department for International Development, DFID, UK
Lenton, Roberto, United Nations Development Programme, UNDP, USA
Segnestam, Mats, Swedish International Development Co-operation Agency, Sida, Sweden

FSG members:

Chappé, Philippe, Ministry of Foreign Affairs, France (alternates with Dutch representative)
Diphoorn, Bert, Ministry of Foreign Affairs (DML/PS), The Netherlands (alternates with French representative)

Hartmann, Armon, Swiss Agency for Development, Switzerland (alternates with German representative)
Helming, Stefan, GTZ, Germany (alternates with Swiss representative)

Ex-officio members:

Ghosh, Gourisankar, UNICEF, Water, Environment and Sanitation, UN ACC, UN Subcommittee on Water Resources, USA
Jonch-Clausen, Torkil, GWP TAC Chair
Matiza-Chiuta, Tabeth, GWP SATAC Chair
Mohtadullah, Khalid, GWP Executive Secretary, Sweden

Permanent observer:

Jamil Al-Alawi, World Water Council, France

Source: Global Water Partnership <http ://www.gwpforum.org/contactsgov.htm>.

Appendix 2

Technical Advisory Committee of the GWP, 2000

Anil Agarwal, Centre for Science and Environment, India

Marian S. delos Angeles, Environmental and Natural Resources Accounting Project, Philippines

Ramesh Batia, Resources and Environment Group, India (left in 2000)

Ivan Chéret, Lyonnaise des Eaux, France

Martin Falkenmark, Stockholm International Water Institute, Sweden

Torkil Jonch-Clausen, VKI Water Quality Institute, Denmark

Mohammed Ait Kadi, Ministère de l'Agriculture et du Dévelopement Rural et des Pêches Maritimes, Morocco

Janusz Kindler, Warsaw University of Technology, Poland (left in 2000)

Judith Rees, London School of Economics, UK

Romana P. de los Reyes, Ateneo de Manilla University, Philippines (left in 2000)

Paul Roberts, Department of Water Affairs and Forestry, South Africa

Peter Rogers, Harvard University, USA

Miguel Solanes, Economic Commission for Latin America and the Caribbean, Chile

Fernando Gonsalez Villareal, Universidad Necional Autonomade Mexico, Mexico (left in 2000)

Albert Wright, The World Bank Resident Mission, Ghana

Source: Global Water Partnership <http ://www.gwpforum.org/contactsgov.htm>.

Appendix 3

WCD, composition of the Second Forum Meeting, April 2000

Affected people's groups

- CODESEN, Coordination for the Senegal River Basin, Senegal
- Cordillera People's Alliance
- Federacion de Indigenas del Estado Boliviar/COICA, Venezuela
- Grand Council of the Cree, Canada
- MAB, Movimento dos Antigos por Barragens, Brazil
- NBA, Narmada Bachao Andolan, India
- SUNGI Development Foundation, Pakistan

Bilateral agencies/ Export Credit Guarantee Agencies

- BMZ, Federal Ministry for Economic Co-operation and Development, Germany
- JBIC, Japan
- NORAD, Norwegian Agency for International Cooperation, Norway
- SDC, Swiss Agency for Development and Cooperation, Switzerland
- Sida, Swedish International Development Agency, Sweden
- US Export–Import Bank, USA

Government agencies

- US Bureau of Reclamation, USA
- LHWP, Lesotho Highlands Water Project, Lesotho
- Ministry of Water Resources, China
- National Water Commission, Mexico
- Ministry of Mahaweli Development, Sri Lanka
- Ministry of Water Resources, India

International associations

- ICID, International Commission for Irrigation & Drainage, New Delhi
- ICOLD, International Commission on Large Dams, South Africa

- IEA, International Energy Agency, Paris
- International Hydropower Association (IHA)
- International Association for Impact Assessments (IAIA)

Multilateral agencies

- ADB, Asian Development Bank, Manila
- AfDB, African Development Bank, Abidjan
- FAO, UN Food and Agriculture Organization, Rome
- IDB, Inter-American Development Bank, Washington, DC, USA
- UNDP, United Nations Development Programme, New York, USA
- UNEP, United Nations Environment Programme, Nairobi
- WB, World Bank, Washington, DC, USA

NGOs

- Berne Declaration, Switzerland
- ENDA, Environmental Development Action, Senegal
- Help the Volga River, Russia
- IRN, International Rivers Network, USA
- ITDG, Intermediate Technology Development Group, UK
- IUCN, The World Conservation Union, Switzerland
- Sobrevivencia – Friends of the Earth, Paraguay
- WWF, World Wide Fund for Nature, Switzerland
- DAWN, Fiji
- Transparency International (TI)
- WEED, Germany
- Swedish Society for Nature Conservation
- Wetlands International, Japan

Private sector firms

- Enron, USA
- Harza Engineering Firm, USA
- Siemens, Germany
- ABB, Switzerland
- Saman Engineering Consultants
- Hopewell Holdings
- Electric Power Development Corporation

Research institutes/resource persons

- Confederacion Hydrografica del Ebro, Spain
- Mekong River Commission, Cambodia

- Volta River Authority, Ghana
- Jordan Valley Authority, Jordan

Utilities

- Electrobas, Brazil
- Hydro-Quebec, Canada
- Nepal Electricity Authority, Nepal
- Mini Hydro Division, Philippines
- Electricité de France, France

Source: World Commission on Dams <http://www.dams.org>.

Appendix 4

Composition of WCD Commissioners, February 1998

Professor Kader Asmal, Chair, Minister for Water Affairs and Forestry, South Africa

Donald J. Blackmore, Chief Executive of the Murray–Darling Basin Commission, Australia

Joji Carino, Executive Secretary of the International Alliance of the Indigenous-Tribal Peoples of the Tropical Forest

José Goldemberg, Professor at the University of São Paolo, Brazil

Laxmi Chand Jain, Vice Chair, former Chairperson of the Industrial Development Services, India, Indian High Commissioner in South Africa

Goran Lindhal, President and CEO of ABB Asea Brown Boveri Ltd, Zurich

Deborah Moore, Senior Scientist at the Environmental Defense Fund, USA

Mehda Patkar, social scientist and founder of the Narmada Bachao Andolan (Struggle to Save the Narmada River) in India, founding member of the National Alliance of People's Movement

Wolfgang Pircher, consultant in the field of hydropower, past President and current Honorary President of the International Commission on Large Dams

Thayer Scudder, Professor of Anthropology at the California Institute of Technology, USA

Shen Guoyi, Director-General of the Department of International Co-operation in the Ministry of Water Resources, People's Republic of China, China

Source: Gleick (1998: 179, Table 6.6).

Appendix 5

Composition of the World Commission on Water for the 21st Century, 2000

Chairman

Ismail Serageldin, Vice President, The World Bank, and Chairman, Consultative group for International Agricultural Research and Global Water Partnership

Honorary members:

HRH The Prince of the Netherlands
Norman Borlaug, Nobel Laureate, USA
Hon. Ingvar Carlsson, Former Prime Minister of Sweden
Jean Dausset, Nobel Laureate, France
Hon. Mikhail Gorbachev, Former President of the Former USSR
Henry Kendall, Nobel Laureate, USA
Hon. Sir Ketumile Masire, Former President of Botswana
Hon. Fidel Rasmos, Former President, The Philippines

Members:

Shahrizaila bin Abdullah, Hon. President, International Commission on Irrigation and Drainage, Malaysia
Anil Agarwal, Director, Centre for Science and the Environment, India
Abdel Latif Al-Hamad, Chairman of the Board, Arab Fund for Economic and Social Development, Kuwait
Kader Asmal, Professor and Chairman of the World Commission on Dams; Minister of Education of South Africa
Asit Biswas, President, Third World Center for Water Management, Mexico City, Mexico
Margaret Catley-Carlson, International Consultant; former President CIDA and Population Council
Gordon Conway, President, The Rockfeller Foundation
Mohamed T. El-Ashry, Chairman and CEO, Global Environmental Facility
Howard Hjort, Former Deputy Director-General, FAO

Enriqué Iglesias, President, Inter-American Development Bank
Yolanda Kakabadse, President, The World Conservation Union
Speciosa Wandira Kazibwe, Vice-President, Uganda
Jessica Matthews, President, Canergie Endowment for International Peace
Robert S. McNamara, Co-Chair, Global Coalition for Africa
Jérôme Monod, Chairman of the Supervisory Board, Suez Lyonnaise des Eaux, France
Peter Rogers, Division of Engineering and Applied Sciences, Harvard University
Maurice Strong, Chairman, Earth Council
Kazuo Takahashi, Director, International Development Research Institute
Wilfried Thalwitz, Former Senior Vice President, The World Bank
José Israel Vargas, Minister of Science and Technology, Brazil, and President, Third World Academy of Sciences, Brazil

Source: Cosgrove and Rijsberman (1999: Annex A).

Appendix 6

Five types of public sector participation arrangements

'Gestion en régie' (local authority control)

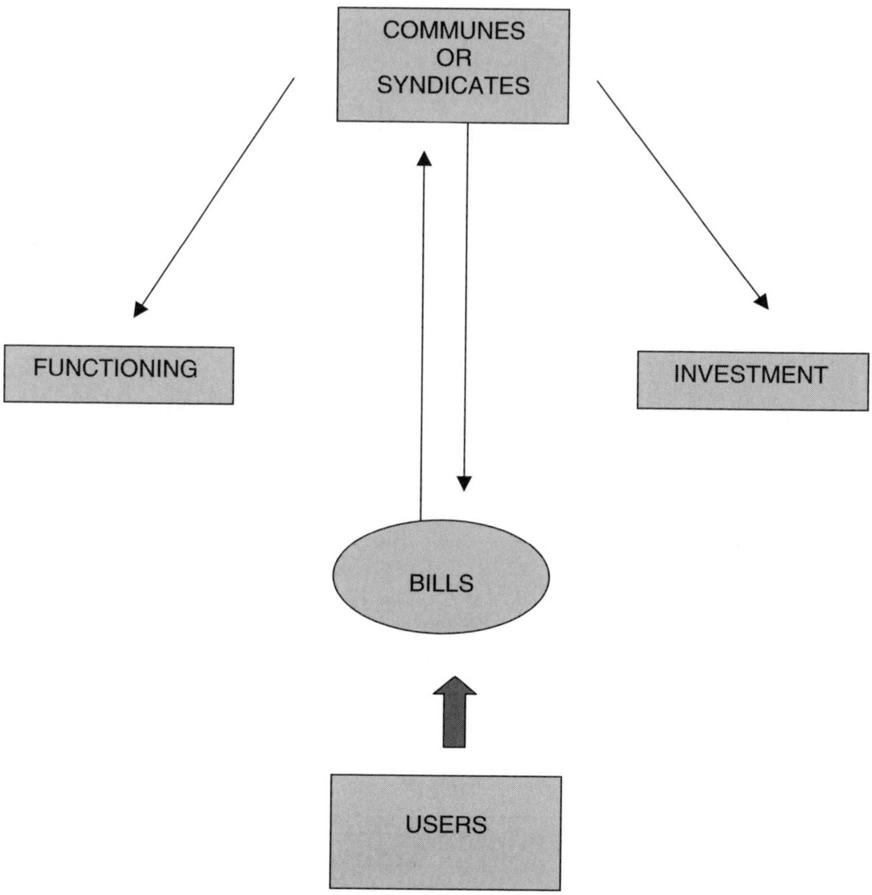

'Gestion en régie avec gérance privée' (Local authority with private management)

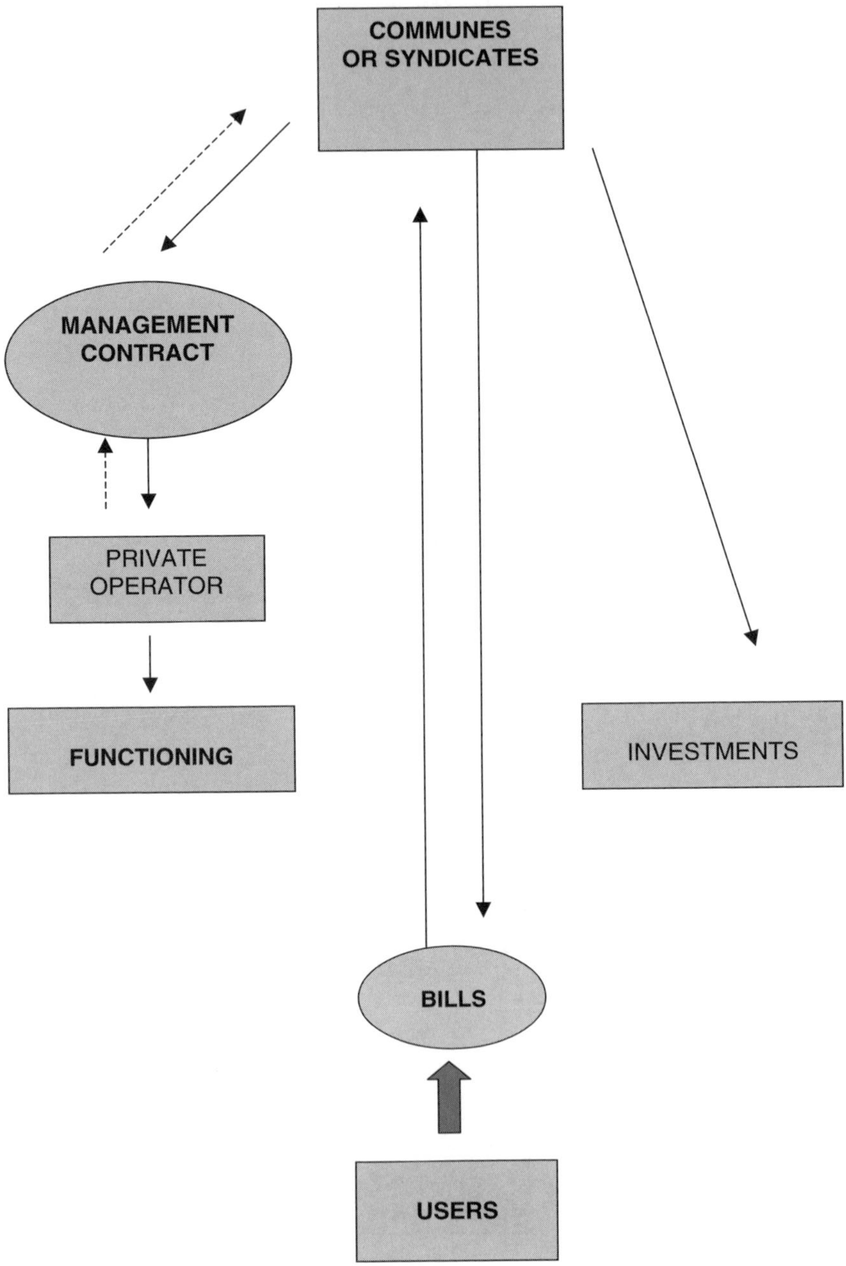

'Gestion déléguée' (1), affermage (lease contract)

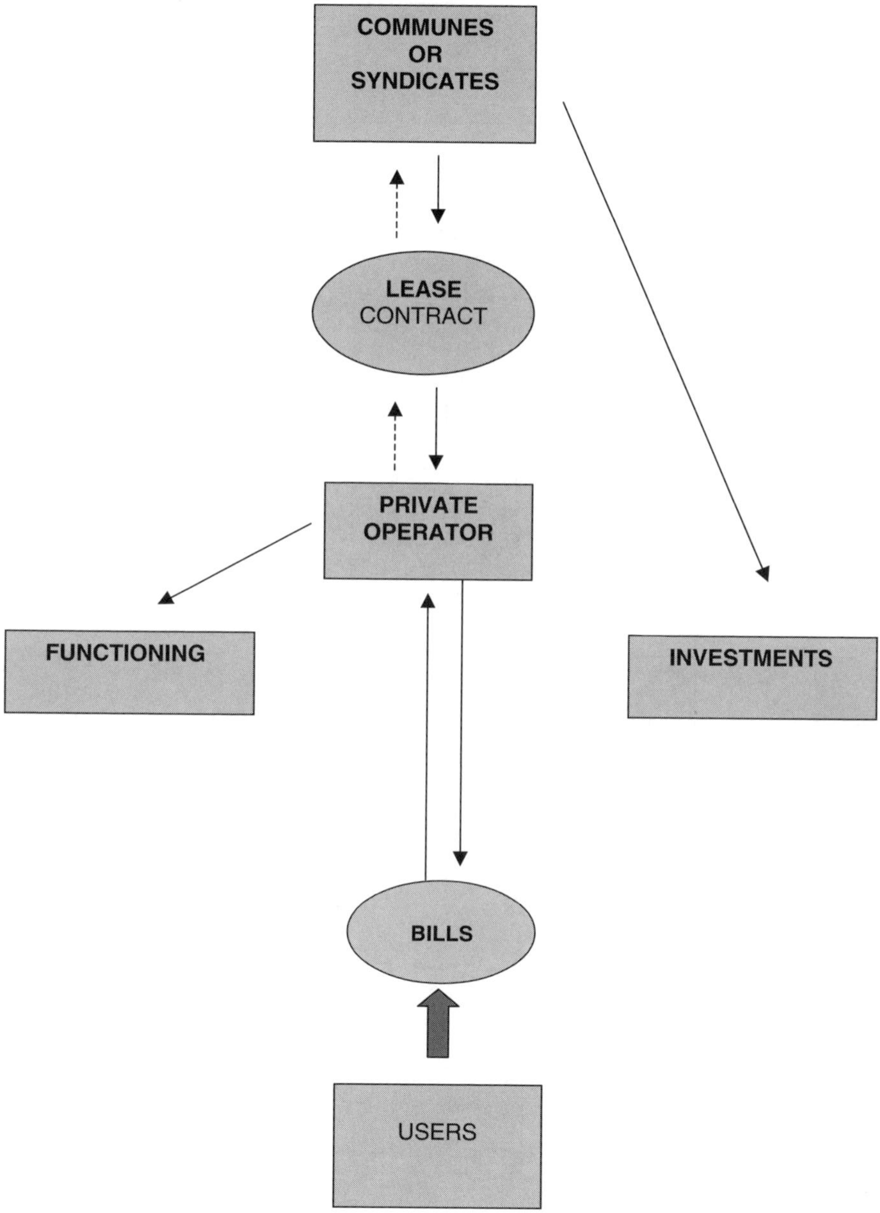

'Gestion déléguée' (2), concession

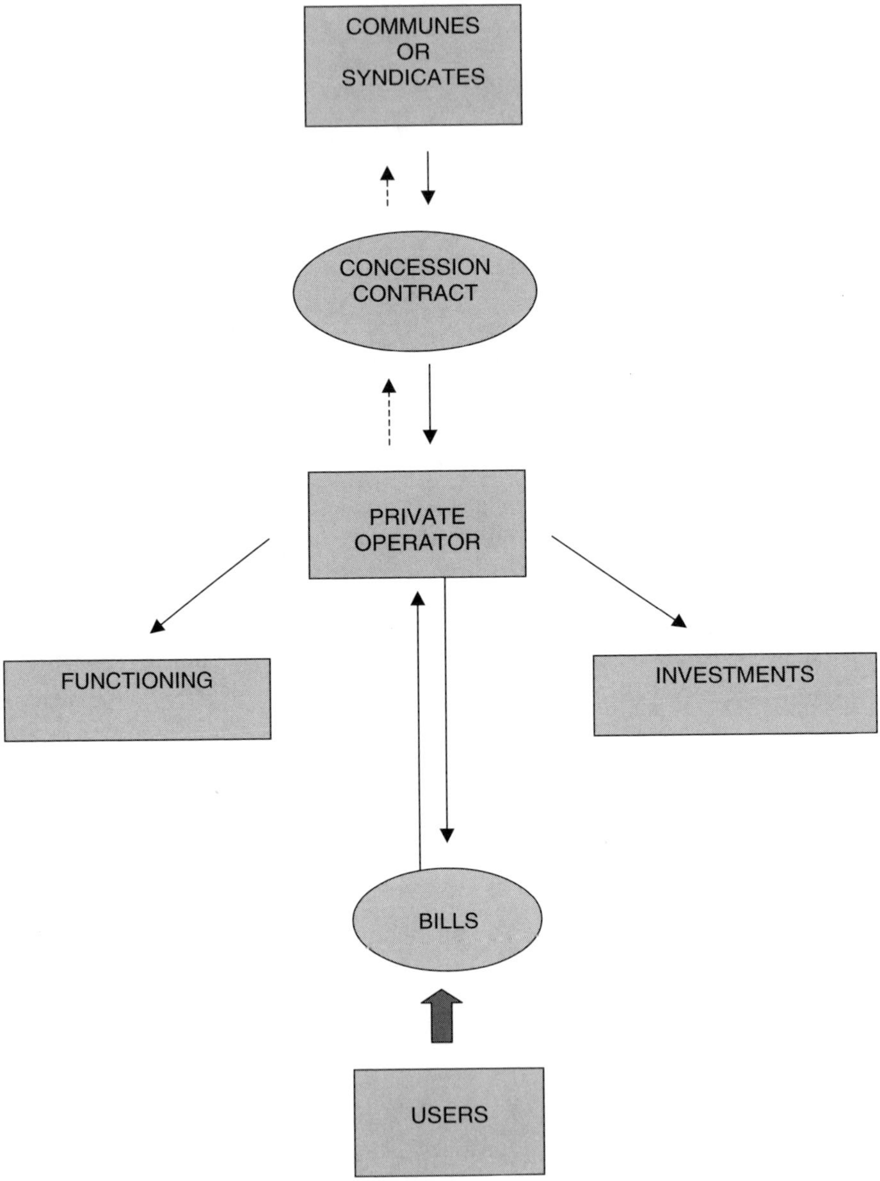

Société d'économie mixte (semi-public company)

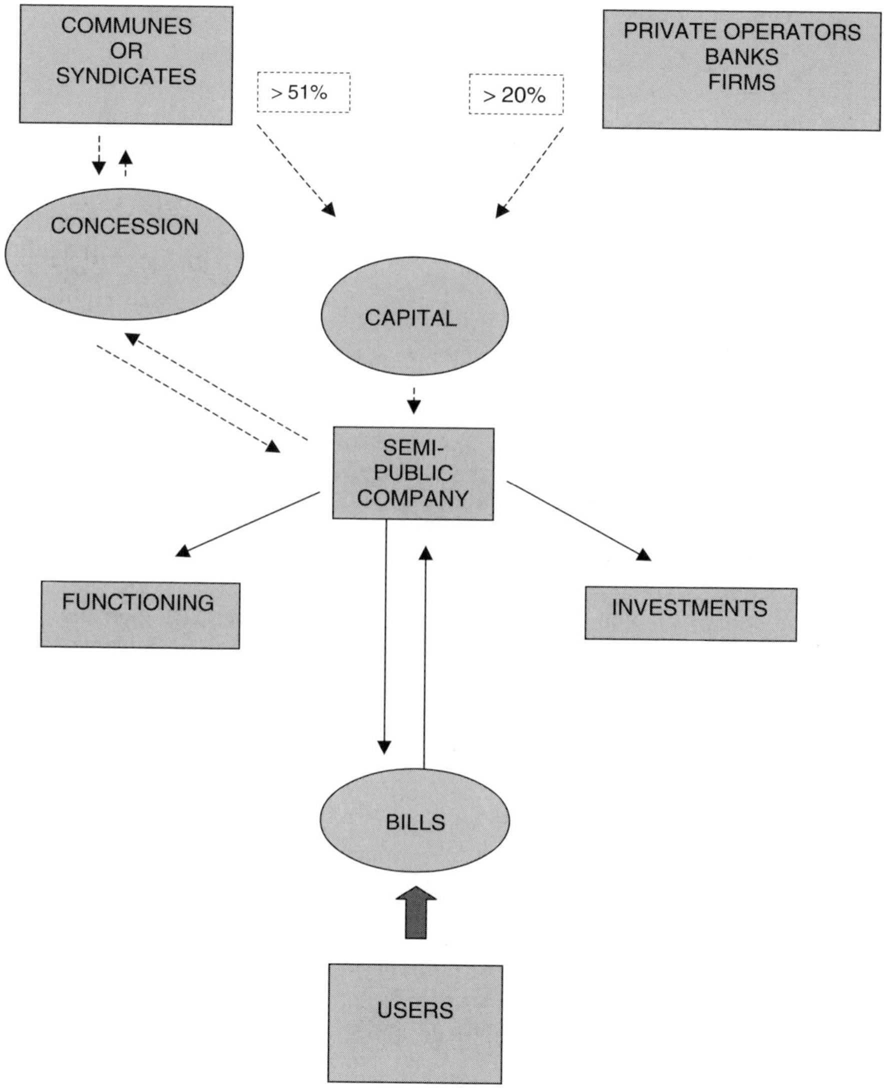

Bibliography

Akinsanya, Adeoye (1984), *Multinationals in a Changing Environment: a Study of Business–Government Relations in the Third World*, New York, Westport, etc.: Praeger.

Alger, Chadwick F. (ed.) (1998), *The Future of the United Nations System: Potential for the Twenty-First Century*, Tokyo: United Nations University Press.

Allouche, Jeremy and Finger, Matthias (2000), 'User Participation in Water Resources Management: The Need to Examine what Exists Behind the Rhetoric', International Water Association Specialised Conference, *Water Management for the 21st Century – Learning from the 20th century experience* (24–25 October).

Allouche, Jeremy and Finger, Matthias (2001), 'Two Reasonings, but One Outcome: The World Bank's Evolving Philosophy in Establishing a "Sustainable Water Resources Management" Policy', *Global Environmental Politics*, 1(2), 42–7 (May).

Anderson, Sarah and Cavanagh, John (2000), *Top 200: The Rise of Corporate Global Power*, Washington, DC: Institute for Policy Studies.

Ascher, Kate (1987), *The Politics of Privatisation: Contracting Out Public Services*, London: Macmillan.

Balassa, Bela A. (1981), 'Structural Adjustment Policies in Developing Countries', *World Bank Staff Working Paper*, 464, Washington, DC: World Bank.

Barlow, Maude (1999), *Blue Gold : The Global Water Crisis and the Commodification of the World's Water Supply*, San Francisco: International Forum on Globalization (IFG) (June).

Barnet, Richard J. and Cavanagh, John (1994), *Global Dreams. Imperial Corporations and the New World Order*, New York: Simon & Schuster.

Barraqué , Bernard (sous la direction de) (1995), *Les Politiques de l'Eau en Europe*, Paris: Editions La Découverte.

Bathia, Ramesh and Falkenmark, Malin (1994), *Water Resource Policies and the Urban Poor: Innovative Approaches and Policy Imperatives*, Washington, DC: UNDP–World Bank Water and Sanitation Program.

Bauby, Pierre (1998), *Reconstruire l'action publique. Services publics, au service de qui?*, Paris: La Découverte et Syros.

Bauer, Carl J. (1998), *1960 – Against the Current: Privatization, Water Markets, and the State in Chile*, Boston: Kluwer Academic.

Baumol, William Jack (1969), *Welfare Economics and the Theory of the State*, London: The London School of Economics and Political Science.

Beecher, Janice A., Dreese, G. Richard and Standord, John D. (1995), *Regulatory Implications of Water and Wastwater Utility Privatization*, Columbus, Ohio: National Regulatory Research Institute.

Beesley, Michael Edwin and Littlechild, Stephen Charles (1991), 'The Regulation of Privatised Monopolies in the United-Kingdom' in C. Velganvski (ed.), *Regulation and the Market*, London: London Institute of Economic Affairs, 29–58.

Benlahcen Tlemcani, Mohamed (1993), 'Le secteur public dans les pays en voie de développement face au defi moins d'Etat', *Politiques et Management Public*, 11(3).

Bishop, Matthew and Thompson, David (1996), 'Privatisation in the UK : Internal Organisation and Productive Efficiency', Chapter 18 in Matthew Bishop, John Kay and Colin Mayer (eds), *Privatization and Economic Performance*, Oxford: Oxford University Press.

Bislev, Sven, Hansen, Hans Krause and Salskov-Iversen, Dorte (2000), *Transnational Discourse Communities: Globalizing Public Management*, Paper for 41st Annual Convention, International Studies Association Conference (Los Angeles, 14–18 March), Panel on Discourses of Globalisation.

Biswas, Asit K. (ed.) (1978), *United Nations Water Conference: Summary and Main Documents*, Oxford, New York, etc.: Pergamon Press.

Bizaguet, Armand (1992), *Le secteur public et les privatisations*, 2nd edn, Paris: Presses Universitaires de France.

Blokland, Maarteen, Braadbaart, Okke and Schwartz, Klaas (eds) (1999), *Private Business, Public Owners, Government Shareholdings in Water Enterprises*, Nieuwegein: Ministry of Housing, Spatial Planning, and the Environment, Water Supply and Sanitation, Collaborative Council.

BMZ (1996), *Stratégie sectorielle: Gestion de l'eau en milieu urbain*, Bonn: BMZ.

Bouin, Olivier and Morrisson, Christian (sous la direction de) (1996), *Les privatisations: un état des lieux*, Paris: Presses de Sciences-Po, 1219–1428.

Boyer, Robert and Drache, Daniel (eds) (1996), *States against Markets: The Limits of Globalization*, London: Routledge.

Briscoe, John (1992), 'Poverty and Water Supply: How to Move Forward', *Finance and Development* (December), 16–19.

Briscoe, John (1993), 'When the Cup is Half Full: Improving Water and Sanitation Services in the Developing World', *Environment*, 35(4).

Briscoe, John (1996), *Water as an Economic Good: The Idea and What it Means in Practice*, World Congress of the International Commission on Irrigation and Drainage, Cairo (September).

Briscoe, John (1997), *Managing Water as an Economic Good: Rule for Reformers*, The World Bank, keynote paper to The International Committee on Irrigation and Drainage Conference on Water as an Economic Good (Oxford, September).

Broclawski, Jean-Pierre, Longueville, Guy and Uhel, Pierre (1986), *Les privatisations à l'étranger: Royaume-Uni, RFA, Italie, Espagne, Japon*, Paris: La Documentation française.

Brook Cowen, Penelope J. (1997a), 'Getting the Private Sector Involved in Water – What to do in the Poorest of Countries?', *Public Policy for the Private Sector*, Note 102, Washington, DC: World Bank (January).

Brook Cowen, Penelope J. (1997b), 'The Private Sector in Water and Sanitation – How to Get Started', *Public Policy for the Private Sector*, Note 126, Washington, DC: World Bank (September).

Brook Cowen, Penelope J. (1999), 'Lessons from the Guinea Water Lease', *Public Policy for the Private Sector*, Note 78, Washington, DC: World Bank (April).

Brook Cowen, Penelope J. and Cowan, Tyler (1998), 'Deregulated Private Water Supply: A Policy Option for Developing Countries', *Cato Journal* (Spring/ Summer).

Brook Cowen, Penelope J. and Tynan, Nicola (1999), 'Reaching the Urban Poor with Private Infrastructure', *Public Policy for the Private Sector*, Note 188, Washington, DC: World Bank (June).

Buckley, Peter J. and Casson, Mark (1976), *The Future of the Multinational Enterprise*, London: Macmillan.

Cabaret, Bernard (1991), 'Les distributeurs d'eau face au grand marché de 1993', *Courants*, 8, 22–8.

Caisse française de développement (1997), *Privatisations et développement* (sous la dir. de Jean-Pierre Barbier and Arnaud Bellamy-Brown), Paris: La Documentation française.

Cans, Roger (1994), *La Bataïlle de l'Eau*, Paris: Le Monde Editions.

Cavanagh, John, Anderson, Sarah and Pike, Jill (1996), 'Behind the Cloak of Benevolence: World Bank and IMF Policies Hurt Workers At Home and Abroad', in Kevin Danaher (ed.), *Corporations Are Gonna Get Your Mama: Globalization and the Downsizing of the American Dream*, Monroe, Maine: Common Courage Press.

Chapman, Ross and Cuthbertson, Sandy (1999), 'Sydney's Water – A Suitable Case for Private Treatment?', *Public Policy for the Private Sector*, Note 80, Washington, DC: World Bank (April).

Chatterjee, Pratap and Finger, Matthias (1994), *The Earth Brokers: Power, Politics and World Development*, London and New York: Routledge.

Chesnais, François (1999), *A Critical View on the Current Consolidation Fever*, May <http://attac.org>.

Chossoudovsky, Michel (1997), *The Globalisation of Poverty. Impacts of IMF and World Bank Reforms*, London: Zed Books.

Cohen, E. (1996), *La tentation hexagonale. La Souveraineté à l'épreuve de la mondialisation*, Paris: Fayard.

Conseil de l'Europe (1994), 'La privatisation d'entreprises et d'activités publiques', Recommandation R (93) 7 adoptée par le Comité des Ministres du Conseil de l'Europe le 18 octobre 1993 et exposé des motifs, Strasbourg: Conseil de l'Europe.

Cook, Paul and Kirkpatrick, Colin (eds) (1988), *Privatisation in Less Developed Countries*, Brighton: Wheatsheaf, New York: St Martin's Press.

Cosgrove, William J. and Rijsberman, Frank R. (eds) (1999), *World Water Vision: Report, World Water Commission's Staff Report*, Marseilles: World Water Council, Version of 17 December.

Cowan, Simon (1994), 'Privatization and Regulation of the Water Industry in England and Wales', Chapter 5 in Matthew Bishop, John Kay and Colin Mayer (eds), *Privatization and Economic Performance*, Oxford: Oxford University Press.

Cox, Helmut (1999), 'Regulation versus Public Property: A Comparative Analysis', *Annals of Public and Competitive Economics*, 70(2).

Crampes, Claude and Estache Antonio (1996), 'Regulating Water Companies: Lessons from the Buenos Aires Concession', *Public Policy for the Private Sector*, Note 91, Washington, DC: World Bank (September).

Culem, Claudy G. (1988), 'The Locational Determinants of Direct Investments among the Industrialized Countries', *European Economic Review*, 32, 885–904.

Cummings, Ronald G., Dinar, Ariel and Olson, Douglas (1996), 'New Evaluation Procedures for a New Generation of Water-Related Projects', *World Bank Technical Paper*, No. 349, Washington, DC: World Bank.

Cutler, Claire (2000), 'Globalisation, Law, and Transnational Corporations: A Deepening of Market Discipline', in Theodore Cohn and Stephen McBride (eds), *Power in the Global Era*, London: Macmillan.

DANIDA (1992), Ministry of Foreign Affairs, *Water Supply and Sanitation*, Copenhagen: DANIDA.

DANIDA (1998), Ministry of Foreign Affairs, *Environmental Issues in Water Resources Management*, Copenhagen: DANIDA.

David, Cristina C. (1999), 'Private Sector Participation in Water Supply and Sanitation: Realising Social and Environmental Objectives in Manila', Chapter 5 in Nick Johnstone and Libby Wood (eds), *Private Sector Participation in Water Supply and Sanitation: Realising Social and Environmental Objectives*, Summary of proceedings of workshop held in London (26–27 November 1998), Draft report to DANIDA, London: International Institute for Environment and Development (August).

Defeuilley, Christophe (1999), 'Competition and Public Service Obligations: Regulatory Rules and Industries' Games', *Annals of Public and Cooperative Economics*, 70(1).

Dehesa, Guillermo de la (1991), *Privatization in Eastern and Central Europe*, Washington, DC: Group of Thirty.

De Luca, Loretta (ed.) (1998), *Labour and Social Dimensions of Privatisation and Restructuring (Public Utilities: Water, Gas, Electricity)*, Geneva: International Labour Organisation.

Deme, Mouhamou (1995), 'Les privatisations, une solution pour l'Afrique?', *Politiques et Management Public*, 13(2).

Demsetz, Harold (1989), *Efficiency, Competition, and Policy, The Organisation of Economic Activity*, II, Oxford: Basil Blackwell.

Department for International Development (DFID) (1997), 'Private Sector Participation in the Water and Sanitation Sector' (Richard Franceys), *DFID Occasional Paper* (3) (July).

Deveaux, Dominique (1998), 'Privatiser, mais comment', *Jeune Afrique*, 1935 (10–16 February 1998).

Dinar, Ariel (ed.) (2000), *The Political Economy of Water Pricing*, New York: Oxford University Press.

Dinar, Ariel, Rosegrant, Mark W. and Meinzen-Dick, Ruth (1997), 'Water Allocation Mechanism: Principles and Examples', *Policy Research Working Paper*, Note 1779, Washington, DC: World Bank (June).

Dinar, Ariel and Subramanian, Ashok (eds) (1997), 'Water Pricing Experiences, an International Perspective', *World Bank Technical Paper*, 386, Washington, DC: World Bank.

Dunning, John H. (1980), 'Towards an Eclectic Theory of International Production', *Journal of International Business Studies*, 11(1), 9–30.

Dunning, John H. (1993a), *The Globalisation of Business: the Challenge of the 1990s*, London, New York: Routledge.

Dunning, John H. (1993b) *Multinational Enterprises and the Global Economy*, Wokingham and Reading, Mass.: Addison-Wesley.

Dunning, John H. (1998), *American Investment in British Manufacturing Industry, Revised and Updated Edition*, London: Routledge.

Dunning, John H. and Hamdani, Khalil A. (1997), *The New Globalism and Developing Countries*, Tokyo, New York and Paris: United Nations University Press.

Durupty, Michel (1988), *Les privatisations en France*, Paris: La Documentation française.

Easter, William K., Rosegrant, Mark W. and Dinar, Ariel (1998), *Markets for Water, Potential and Performance*, Boston, Dordrecht and London: Kluwer Academic Publishers.

Eberlein, Burkard (1999), 'L'Etat régulateur en Europe', *Revue Française de Science Politique*, 49(2), 205–30.

Ecole Nationale d'Administration (ENA) (1997), 'Promotion March Bloch (1995–1997)', *Services Publics comparés en Europe: exception française, exigence européenne*, Tome 1, Paris: La Documentation Française, Chapitre IV: Distribution de l'Eau et autre Services Urbains, 247–319.

Euijune, Kim (1998), 'Spatial Integration of Urban Water Services and Economies of Scale', *Rurds*, 10(1).

Finger, Matthias and Lobina, Emanuele (1999), 'Managing Globalisation in Public Utilities: Public Service Transnational Corporations and the Case of the Global Water Industry', *Cahier de l'IDHEAP*, 177, Chavannes-près-Renens: IDHEAP (Institut de Hautes Etudes en Administration Publique), UER: Management des entreprises publiques.

Finger, Matthias and Ruchet, Bérangère (2000), 'The Transformation of International Public Organisations: The Case of UNCTAD', *Working Paper de l'IDHEAP*, 14, Chavannes-près-Renens: IDHEAP, UER: Management des Entreprises Publiques.

Finger, Matthias and Tamiotti, Ludivine (eds) (2001), *The Multi-Governance of Water Through Four Case Studies*, Report to IUCN, Gland : IUCN.

Food and Agriculture Organization (FAO) (1993), *The State of Food and Agriculture 1993*, Rome: FAO.

Food and Agriculture Organization (FAO) (1996), 'Guidelines for Planning Irrigation and Drainage Investment Projects', *FAO Technical Paper*, 11, Rome: FAO.

Franceys, Richard (1997), 'Private Sector Participation in the Water and Sanitation Sector', Loughborough University and IHE, Delft (July), *Water Resources Occasional Paper*, 3.

Frederiksen Harald D. (1992), 'Discussion of Some Misconceptions about Water-Use Efficiency and Effectiveness', *World Bank Technical Paper*, 181, Washington, DC: World Bank.

Frederiksen, Harald D., Berkoff, Jeremy and Barber, William (1993), *Water Resources Management in Asia*, Washington, DC: World Bank.

Friedman, Milton (1953), *Essays in Positive Economics*, Chicago, London: University of Chicago Press.

Fundanga, Caleb M. and Mwaba, Andrew (1998), 'Privatization of Public Enterprises in Zambia, An Evaluation of the Policies, Procedures and Experiences', *Economic Research Papers*, 35, Abidjan: African Development Bank.

Gibbons, Diana C. (1987), *The Economic Value of Water*, Washington, DC: Resources for the Future.

Giddens, Anthony (1990), *The Consequences of Modernity*, Cambridge: Polity Press.
Gleick, Peter H. (1998), *The World's Water: The Biennial Report on Freshwater Resources, 1998–1999*, Washington, DC: Island Press.
Global Water Partnership (GWP) (1998), *Global Water Partnership 1999: a Strategic Plan*, Stockholm: GWP Publications.
Global Water Partnership (GWP) (1999), *GWP, Financial Support Group, Meeting II*, Stockholm: GWP Publications (21 March).
Great Britain. Monopolies and Mergers Commission (MMC) (1995), *Lyonnaise des Eaux SA and Northumbrian Water Group PLC: a Report on the Merger Situation*, Monopolies and Mergers Commission; presented to Parliament by the Secretary of State for Trade and Industry by Command of Her Majesty, July, London: HMSO.
Great Britain (1996), *Water, Increasing Customer Choice: Competition in the Water and Sewerage Industries: the Government's Proposals*, Department of the Environment, Welsh Office.
Grosse, Robert and Behrman, Jack N. (1992), 'Theory in International Business', *Transnational Corporations*, 1(1), 93–121.
Haarmeyer, David (1992), 'Privatizing Infrastructure: Options for Municipal Water Supply Systems', Executive Summary, *Policy Study*, 151, Los Angeles: Reason Foundation (October).
Haarmeyer, David and Mody, Ashoka (1997), 'Capitaux privés et gestion de l'eau', *Finances et développement* (March).
Haarmeyer, David and Mody, Ashoka (1998), 'Financing Water and Sanitation Projects – The Unique Risks', *Public Policy for the Private Sector*, Note 151, Washington, DC: World Bank (September).
Habermas, Jürgen (1980), *Legitimation Crisis*, London: Heinemann.
Haigh, Nigel (1996), *The Manual of Environmental Policy: The EC and Britain*, London: Catermill Publishing.
Hall, David (1998a), 'Public Enterprise in Europe' in G. Hotham (ed.), *Freedom with Responsibility – Can We Unshackle Public Enterprise?*, London: International Institute for Public Policy Research.
Hall, David (1998b), 'Restructuring and Privatisation in the Public Utilities – Europe', in L. De Luca (ed.), *Labour and Social Dimensions of Privatisation and Restructuring – (Public Utilities: Water, Gas and Electricity)*, Geneva: ILO.
Haller, Laurence (1998), *Evaluations of the Dublin Principles in the Urban Water and Sanitation Sector* (September), MSc in Environmental Assessment and Evaluation, LSE, unpublished.
Hanson, Todd and Oliveira, Letitia (1997), 'International Finance Corporation', *Environment Matters*, Washington, DC: World Bank (Fall), 32–5.
Hearne, Robert R. and Easter, William (1995), *Water Allocation and Water Markets: An Analysis of Gains-from-Trade in Chile*, Washington, DC: World Bank.
Hearne, Robert R. and Trava, José L. (1997), 'Water Markets in Mexico', *Discussion Paper*, 97–01, London: International Institute for Environment and Development (IEED) (April).
Held, David (1995), *Democracy and Global Order: From the Modern State to Cosmopolitan Governance*, Stanford: Stanford University Press.
Hirst, Paul and Thomson, Grahame (1996), *Globalization in Question*, Cambridge: Cambridge University Press.

Holing, Dwight (1997) 'California's Hot Water Market: How Water Transfers are Protecting the Economy and the Environment', *Tomorrow*, 5(VII) (September–October), 6.

Hollingsworth, Rogers J. and Boyer, Robert (eds) (1997), *Contemporary Capitalism: The Embeddeness of Institutions*, Cambridge: Cambridge University Press.

Hymer, Stephen H. (1976), *The International Operations of National Firms: A Study of Direct Foreign Investment*, Cambridge, Mass.: MIT Press.

Idelovitch, Emanuel and Ringskog, Klas (1997), *Private Sector Participation in Water Supply and Sanitation in Latin America*, Washington, DC: World Bank, Directions in Development.

International Institute for Environment and Development (1999), *Community Water Management, pla Notes 35: Participatory Learning and Action*, London: International Institute for Environment and Development, Sustainable Agriculture and Rural Livelihoods Programme (June).

Izaguirre, Ada Karina and Rao, Geetha (2000), 'Private Infrastructure', *Private Sector and Infrastructure Network Note 215* (September).

Johnstone, Nick and Wood, Libby (eds) (1999), *Private Sector Participation in Water Supply and Sanitation: Realising Social and Environmental Objectives*, Summary of proceedings of workshop held in London (26–27 November 1998), Draft report to DANIDA, London: International Institute for Environment and Development (August).

Jones, Julian (1997), *Urban Water Supplies for Developing Countries: The French Approach*, Institutional Development Series, The Water, Engineering and Development Centre (WEDC), Loughborough University.

Kaczmarek, Bernard (1997), 'La politique communautaire de l'eau', *Aménagement et nature* (March).

Katakura, Yoko and Bakalian, Alexander (1998), *PROSANEAR, People, Poverty and Pipes, A Programme of Community Participation and Low Cost Technology Bringing Water and Sanitation to Brazil's Urban Poor*, Washington, DC: UNDP World Bank, Water and Sanitation Program (September).

Kerf, Michel and Smith, Warrick (1996), 'Privatising Africa's Infrastructure: Promise and Challenge', *World Bank Technical Paper*, 337, Washington, DC: World Bank.

Kikeri, Sunita, Nellis, John and Mary, Shirley (1992), *Privatization: The Lessons of Experience*, Washington, DC: World Bank.

Kirmani, Syed and Rangeley, Robert (1994), 'International Inland Waters, Concept for a More Active World Bank Role', *World Bank Technical Paper*, 239, Washington, DC: World Bank.

Kiser, Edgar and Baker, Kathryn (1994), 'Could Privatisation Increase the Efficiency of Tax Administration in Less Developed Countries?', *Policy Studies Journal*, 22(3).

Klein, Michael (1996), 'Economic Regulation of Water Companies', *World Bank Policy Research Working Paper*, 1649, Washington, DC: World Bank (September).

Klein, Michael and Irwin, Timothy (1996), 'Regulating Water Companies', *Public Policy for the Private Sector*, Note 77, Washington, DC: World Bank (May).

Komives, Kristin (1999), 'Designing Pro-Poor Water and Sewer Concessions', *Policy Research Working Paper*, 2243, Washington, DC: World Bank (November).

Komives, Kristin and Brook Cowen, Penelope J. (1999), 'Expanding Water and Sanitation Services to Low Income Households, The Case of the La-Paz–El Alto

Concession', *Public Policy for the Private Sector*, Note 188, Washington, DC: World Bank (April).

Korten, David C. (1996), *When Corporations Rule the World*, London: Earthscan.

Kovar, Robert and Denys, Simon (sous la direction de) (1998), *Service Public et communauté européenne: entre l'interêt général et le marché, Tome 1: Introductions et approche sectorielle*, Acte du colloque de Strasbourg (17–19 October 1996), Paris: La Documentation Française.

Kumar, Ritu and Young, Carlos (1996), 'Economic Policies for Sustainable Water Use in Thailand', *CREED Working Paper, Series 4*, London: International Institute for Environment and Development (IIED) (May).

Lane, Jan-Erik (1985), *State and the Market: the Politics of the Public and the Private*, London and Beverly Hills: Sage.

Larry, Simpson and Ringskog, Klas (1997), *Water Markets in the Americas*, Directions in Development, Washington, DC: World Bank.

Lavigne, Jean-Claude (1995), 'Deux expériences africaines, Côte d'Ivoire, Guinée', in Dominique Lorrain (sous la direction de), *Gestions Urbaines de l'Eau*, Paris: Economica.

Lavigne, Jean-Claude (1999), 'L'expérience ivoirienne', in *Retour d'expériences, Six cas de gestion déléguée à l'étranger* (Groupe Services urbains et développement durable), Paris: Institut des Sciences et des Techniques de l'Equipement et de l'Environement pour le Développement (ISTED) (May).

Le Duc, Michel (1995), *Services publics de réseau et Europe, Les exemples de l'Allemagne, de l'Espagne, de l'Italie et du Royaume Uni (fer, route, air et eau-assainissement)*, Chapter 4, *Les réseaux d'eau et d'assainissement*, Paris: La Documentation Française.

Lee, Terence R. (2000), *Water Management in the 21st Century: The Allocation Imperative*, Northampton, Mass.: Edward Elgar.

Le Grand, Julian and Robinson, Ray (eds) (1985), *Privatisation and the Welfare State*, London and Boston: George Allen & Unwin.

Le Moigne, Guy *et al.* (eds) (1994), *A Guide to the Formulation of Water Resources Strategy*, Washington, DC: World Bank.

Le Moigne, Guy *et al.* (eds) (1994), *Water Policy and Water Markets* (selected papers and proceedings from the World Bank's Ninth Annual Irrigation and Drainage Seminar, Annapolis, Maryland, 8–10 December 1992), Washington, DC: World Bank.

Lévêque, François (1998), *Economie de la réglémentation*, Paris: Editions de la Découverte.

Levitt, Theodore (1985), 'The Globalisation of Markets' in A. M. Kantrow, *Sunrise ... Sunset: Challenging the Myth of Industrial Obsolescence*, New York: John Wiley, 53–68.

Liebenthal, Andres *et al.* (1996), *The World Bank's Experience with Large Dams: A Preliminary Review of Impacts*. Washington, DC: Operations Evaluation Department, The World Bank.

Linares, Carlos and Rosensweig, Fred (1999), 'Decentralisation in El Salvador', Washington, DC: Environmental Health Project USAID, *Activity Report*, 64 (May).

Lipschutz, Ronnie D. (1996), *Global Civil Society and Global Environmental Governance*, Albany: SUNY Press.

Lorrain, Dominique (1994), 'L'ogipole competitif/la régulation des réseaux techniques urbains', *Annales des Mines* (October–November).

Lunn, John (1980), 'Determinants of US Direct Investments in the EEC', *European Economic Review (*13 January), 93–101.

Lyonnaise des Eaux (1998), *Alternative Solutions for Water Supply and Sanitation in Areas with Limited Financial Resources*, Paris: Lyonnaise des Eaux (September).

MacAvoy, Paul W. (ed) (1995), *Deregulation and Privatization in the United States*, Edinburgh: Edinburgh University Press.

MacLean, Mairi (1991), *French Enterprise and the Challenge of the British Water Industry: Water Without Frontiers*, Aldershot and Brookfield, VT: Avebury.

Maloney, William A. and Richardson, Jeremy (1995), *Managing Policy Change in Britain: the Politics of Water*, Edinburgh: Edinburgh University Press.

Marino, Manuel, Stein, John and Wulff, Francisco (1998), 'Management Contracts and Water Utilities: The Case of Monagas in Venezuela', *Public Policy for the Private Sector*, Note 166, Washington, DC: World Bank (December).

Martinand, Claude (1995), *La régulation des services publics*, Rapport de l'atelier 'Régulation' du Groupe 'Réseaux2010', Paris: ASPE.

Marty, Frank (1997), 'International River Management – The Political Determinants of Success and Failure, Introduction and Research Design', *Studien zur Politikwissenschaft*, 305 (December).

Mathys, Alain (1999), *Improving Cost Efficiency of Rural Water Supply Projects: Findings, Conclusions, and Recommendations of a Regional Cost Study*, Field Note 3, La Paz: UNDP/World Bank Andean Region (April).

McCully, Patrick (1996), *Silenced Rivers: The Ecology and Politics of Large Dams*, London: Zed Books.

McCully, Patrick (1997a), *A Critique of the 'World Bank's Experience with Large Dams: A Preliminary Review of Impacts'*, Berkeley: International Rivers Network.

McCully, Patrick (1997b), 'Tackling Down Bad Dams', *World Rivers Review*, 12(4) (August).

McGrew, Anthony and Lewis, Paul *et al.* (1992), *Globalisation and the Nation States*, Cambridge: Polity Press.

Merkel, W.I. (2000), 'Germany: Preparing for the 21st Century', *Water 21: Magazine of the International Water Association* (October), IWA Publishing.

Miloradov, Milorad and Marjanovic, Prvoslav (1998), *Guidelines for Conducting Water Resources Assessment*, Paris: UNESCO.

Modelski, George (1972), *Principles of World Politics*, New York: Free Press.

Moraru-de Loë, L. and Mitchell, B. (1993), 'Public–Private Partnerships: Water and Wastewater Services – France', *Water International*, 18(3), 137–46.

Morris, John R. (1990), 'Pricing for Water Conservation', *Contemporary Policy Issues*, 8, 79–91.

Morris, John R. (1991), 'Water Conservation Progress in Denver', *Contemporary Policy Issues*, 11, 35–45.

Munasinghe, Mohan (1992), *Water Supply and Environmental Management: Developing World Applications*, Boulder and San Francisco, etc.: Westview Press.

Musa, Asad *et al.* (1999), *Management of Water Resources: Bulk Water Pricing in Brazil*, Washington, DC: World Bank.

Nankani, Helen (1997), 'Testing the Waters – A Phased Approach to a Water Concession in Trinidad and Tobago', *Public Policy for the Private Sector*, Note 103, Washington, DC: World Bank (January).

Neal, Kathy, Maloney, Patrick, Marson, Jonas and Tamer, Francis (1996), 'Restructuring America's Water Industry: Comparing Investor Owned and Government Owned Water Systems', Los Angeles: Reason Foundation, *Policy Study*, 200 (January).

Netherlands Development Assistance, Ministry of Foreign Affairs (1998), *Water Supply and Sanitation in Developing Countries*, The Hague: NEDA (February).

Newbury, Evithan (2000), *Privatization, Restructuring, and Regulation of Network Utilities*, Cambridge, Mass.: MIT Press.

ODA (1996), *Water and NGOs, Workshop Organised by WEDC on Behalf of ODA*, London: ODA (10 June).

Offe, Claus (1984), *Contradictions of the Welfare State*, London and Sydney, etc.: Hutchinson.

Organisation de coopération et de développement économiques (OCDE) (1996), *La privatisation en Asie, Europe et Amérique latine*, Paris: OECD.

Organisation de coopération et de développement économiques (OCDE) (1998), *Gestion de l'Eau: performance et défi*, Paris: OECD.

Organisation for Economic Cooperation and Development (OECD) (1972), *Water Management: Basic Issues*, Paris: OECD.

Organisation for Economic Cooperation and Development (OECD) (1987), *Pricing of Water Services*, Paris: OECD.

Organisation for Economic Cooperation and Development (OECD) (1989), *Water Resource Management: Integrated Policies*, Paris: OECD.

Organisation for Economic Cooperation and Development (OECD) (1994), 'New Ways of Managing Infrastructure Provision', *Public Management Occasional Paper*, 6, Paris: OECD.

Organisation for Economic Cooperation and Development (OECD) (1997), *Water Subsidies and the Environment*, Document OECD/GD (97) 220, Paris: OECD.

Organisation for Economic Cooperation and Development (OECD) (1998), *Globalisation and the Environment, Perspectives From OECD and Dynamic Non-Members Economics*, OECD Proceedings, Paris: OECD.

Organisation for Economic Cooperation and Development (OECD) (1999), *The Price of Water, Trends in OECD*, Paris: OECD.

Organisation Météorologique Mondiale (1997), *Inventaire exhaustif des ressources mondiales en eau douce*, Genève: OMM.

Perry, C. J., Rock, Michael and Seckler, D. (1997), 'Water as an Economic Good : A Solution or a Problem?', International Irrigation Management Institute, *Research Report*, 14.

Petrella, Ricardo (1996), 'Globalization and Internationalization: The Dynamics of the Emerging World Order', in Robert Boyer and Daniel Drache (eds), *States Against Markets: The Limits of Globalization*, London: Routledge.

Petrella, Ricardo (1999), *Le Manifeste de l'eau. Pour un Contrat mondial*, Lausanne: Editions Page Deux.

Petrella, Ricardo (2001), *The Water Manifesto: Arguments for a World Water Contract*, London: Zed Books.

Pigram, John J. (2000), 'Equilibrer le Nord et le Sud', *L'eau au XXIe siècle, De la Vision à l'Action*, Paris: Editions Futuribles (March).

Postel, Sandra (1992), *Last Oasis: Facing Water Scarcity*, New York and London: W. W. Norton.

Postel, Sandra, Daily, G. C. and Ehrlich, P. R. (1996), ' Human Appropriation of Renewable Fresh Water', *Science*, 271, 785–8.

Pottier, C. (1991), 'Les Groupes européens à la recherche d'une taille mondiale', in J. P. Gilly (ed.), *L'Europe industrielle horizon 1993 – 1 Les groupes et l'intégration européenne*, Paris: La Documentation française.

Princen, Thomas and Finger, Matthias (eds) (1994), *Environmental NGOs in World Politics: Linking the Local to the Global*, London: Routledge.

Prokopenko, Joseph (ed.) (1995), *Management for Privatization: Lessons from Industry and Public Service*, Geneva: International Labour Office.

Public Services Privatisation Research Unit (PSPRU) (1996), *The Privatisation Network – the Multinationals' Bid for Public Services*, London: Public Services International.

Raghavan, Chakravarthi (1997a), 'What is Globalisation?', *Third World Resurgence*, 74, 11–14.

Raghavan, Chakravarthi (1997b), 'MIA or MAI: Global Welfare Rule for TNCs', *Third World Resurgence*, 75, 29–32.

Ramanadham, Venkata Vemuri (1988), *Privatisation in the UK*, London and New York: Routledge.

Ramanadham, Venkata Vemuri (ed.) (1993), *Constraints and Impacts of Privatization*, London and New York: Routledge.

Ramanadham, Venkata Vemuri (ed.) (1994), *Privatization and After: Monitoring and Regulation*, London and New York: Routledge.

Rees, Judith (1997), *Interpreting and Applying the Dublin Principles*, Paper for Technical Advisory Committee, Global Water Partnership.

Rees, Judith (1998), 'Regulation and Private Participation in the Water and Sanitation Sector', *GWP TAC Background Papers*, 1, Sweden: GWP Publications.

Régines, Matthew (2000), *La question des exportations Canadiennes d'eau douce en vrae vers les Etats Unis dans le contexte de l'ALENA*, Graduate Institute of International Studies, Geneva, unpublished.

Ríos Brehm, Monica and Quiroz, Jorge (1995), *The Market for Water Rights in Chile: Major Issues*, Washington, DC: World Bank.

Rivera, Daniel (1996), *Private Sector Participation in the Water Supply and Wastewater Sector: Lessons from Six Developing Countries*, Washington, DC: World Bank, Directions in Development.

Robertson, Roland (1992), *Globalization: Social Theory and Global Culture*, London and Thousand Oaks, Ca: Sage.

Roger, Neil (1999), 'Recent Trends in Private Participation in Infrastructure', *Public Policy for the Private Sector*, Note 196, Washington, DC: World Bank (September).

Rogers, Peter, Bhatia, Ramesh and Huber, Annette (1998), 'Water as a Social and Economic Good: How to Put the Principle into Practice', paper for TAC of the GWP in Namibia, *GWP TAC Background Paper*, 2, Stockholm: GWP Publications.

Rondinelli, Dennis A. (1991), 'Decentralizing Water Supply Services in Developing Countries: Factors Affecting the Success of Community Management', *Public Administration and Development*, 11, 415–30.

Rosenau, James N. (1997), *Along the Domestic–Foreign Frontier: Exploring Governance in a Turbulent World*, Cambridge: Cambridge University Press.

Rosenau, James N. and Durfee, Mary (1995), *Thinking Theory Thoroughly: Coherent Approaches to an Incoherent World*, Boulder, Col., San Francisco: Westview Press.

Rosensweig, Fred and Perez, Eduardo (1999), 'Decentralisation in Central America and the Dominican Republic', *Activity Report*, 76, Washington, DC: Environmental Health Project USAID (May).

Rossi, J. (1997), 'Le principali caratteristiche del sistema francese di gestiore dei servici idrici', in L. Guiffant and M. Merelli (eds), *La riforma dei servici idrici in Italia – Riflessioni e spunti comparativi*, Milan: EGEA, SpA.

Roth, Gabriel (1987), *The Private Provision of Public Services in Developing Countries*, New York: Oxford University Press, published for the Economic Development Institute, World Bank.

Rudolph, Karl-Ulrich (2000a), 'Ten Years After – Experiences from BOX and PPP's in the German Water Sector and Lessons for the Future', International Water Association Specialised Conference, *Water Management for the 21st Century – Learning from the 20th Century Experience*, Berlin, 24–25 October.

Rudolph, Karl-Ulrich (2000b), *Water PSP Experiences in Germany – No Unique Model for the World Market, but Fair Traded Tailor Made Solutions*, unpublished.

Rugman, Alan M. (1996), *The Theory of Multinational Enterprise*, Cheltenham: Edward Elgar.

Rugman, Alan (1999), 'Forty Years of the Theory of the Transnational Corporation', *Transnational Corporations*, 8(2) (August), 51–70.

Rund, Andun (1998), 'Transnational Corporations: Conflict with Policies of Less Developed Countries?', paper presented at the 39th Annual Convention, International Studies Association, March, 1998; Columbia International Affairs online conference proceedings.

Saade, Lilian, Blokland, Maarten and Brikke, François (2000), 'Institutional Needs: Critical Aspects and Opportunities in the Water Supply and Sanitation Sector for the Next Decades', Geneva: Water Supply and Sanitation Council, Vision 21: Water for People: A Shared Vision for Hygiene, Sanitation and Water Supply.

Saghir, Jamal, Sherwood, Elisabeth and Macoun, Andrew (1999), 'Management Contracts in Water and Sanitation – Gaza's Experience', *Public Policy for the Private Sector*, Note 177, Washington, DC: World Bank (April).

Salman Salman M. A. (ed.) (1999), 'Groundwater: Legal and Policy Perspectives: Proceedings of a World Bank Seminar', *World Bank Technical Paper*, 456, Washington, DC: World Bank.

Salman Salman M. A. and Boisson de Chazournes, Laurence (eds) (1999), 'International Watercourses: Enhancing Cooperation and Managing Conflict', *World Bank Technical Paper*, 414F, Washington, DC: World Bank.

Sandouly, Patrick (1998), 'Privatiser mais comment?', *Jeune Afrique*, 1935 (10–16 February), 74–83.

Savedoff, William and Spiller, Pablo (eds) (1999), *Spilled Water: Institutional Commitment in the Provision of Water*, Washington, DC: Inter-American Development Bank.

Schofield, Richard and Shaoul, Jean (1997), 'Regulating the Water Industry, Swimming Against the Tide or Going Through the Motions?', *The Ecologist*, 27(1).

Serageldin, Ismail (1994), *Water Supply, Sanitation, and Environmental Sustainability: The Financing Challenge*, Washington, DC: World Bank.

Serageldin, Ismail (1995a), *Towards Sustainable Management of Water Resources*, Directions in Development, Washington, DC: World Bank.

Serageldin, Ismail (1995b), 'Water Resources Management: a New Policy for a Sustainable Future', *Water International*, 20, 15–21.

Serageldin, Ismail (1995c), 'Water Resources Management: a New Policy for a Sustainable Future', *International Journal of Water Resources Development*, 3, (11).

Serageldin, Ismail (1996) 'Comment résoudre la crise de l'eau', *Notre Planète*, 8(3), 4–7.

Serageldin, Ismail and Steers, Andrew (eds) (1994), *Making Development Sustainable: From Concepts to Action.* Environmentally Sustainable Development Occasional Papers Series, 2, Washington, DC: World Bank.

Sharma, Narendra P. *et al.* (1996), *African Water Resources: Challenges and Opportunities for Sustainable Development*, Washington, DC: World Bank.

SIDA/IRC (1994), *Towards Better Water Resources Management: A Catalogue of Policies and Strategies of External Support Agencies*, Reference Series 10 (J. T. Viisscher and M. Soresson), Sweden: SIDA.

Silvia, Gisele, Tynan, Nicola and Yilmaz, Yesin (1998), 'Private Participation in the Water and Sewerage Sector – Recent Trends', *Public Policy for the Private Sector*, Note 147, Washington, DC: World Bank (August).

Smith, Stephan and Glaser, Antoine (1992), *Ces Messieurs Afrique*, Paris: Calman Levy.

Smith, Zachary (1992), 'Water and Federalism in the 1990s', *Publius, The Journal of Federalism*, 22(3), 129–39.

Société de distribution d'eau de Côte d'Ivoire (SODECI) (1996), *Rapport annuel, exercise 1996*, Abidjan: SODECI.

Solo, Tova Maria (1998), 'Competition in Water and Sanitation, the Role of Small Scale Entrepreneurs', *Public Policy for the Private Sector*, Note 165, Washington, DC: World Bank (December).

Solo, Tova Maria (1999), 'Small Scale Entrepreneurs in the Urban and Sanitation Market', *Environment and Urbanization*, 11(1) (April).

Somma, Mark (1997), 'Institutions, Ideology, and the Tragedy of the Commons: West Texas Groundwater Policy', *Publius, The Journal of Federalism*, 27(1) 1–13.

South Centre (1997), *For a Strong and Democratic United Nations. A South Perspective on UN Reform*, London: Zed Books.

Spero, Joan Edelman and Hart, Jeffrey A. (1997), *The Politics of International Economic Relations*, Fifth Edition, London and New York: Routledge.

Spulber, Nicolas and Sabbaghi, Asghar (1998), *Economics of Water Resources: From Regulation to Privatization*, 2nd edn, Boston: Kluwer Academic.

Strange, Susan (1988), *States and Markets. An Introduction to International Political Economy*, London: Pinter Publishers.

Strange, Susan (1996), *The Retreat of State: The Diffusion of Power in the World Economy*, Cambridge: Cambridge University Press.

Swann, Dennis (1988), *The Retreat of the State: Deregulation and Privatization in the UK and US*, Ann Arbor: University of Michigan Press.

Tharun, Günter *et al.* (eds) (1995), *Experiences in the Development of Small-Scale Water Resources in Rural Areas: Proceedings of the International Symposium on Development of Small-Scale Water Resources in Rural Areas*(Bangkok and Khon Kaen, 21–25 May 1990), Thailand: Carl Duisberg Gesellschaft–South East Asia Program Office.

Thatcher, Margaret (1993), *The Downing Street Years: 1979–1990*, New York: HarperCollins, 682.

Thobani, Mateen (1995), 'Tradable Property Rights to Water : How to Improve Water Use and Resolve Water Conflicts', *Public Policy for the Private Sector*, Note 34, Washington, DC: World Bank (February).

Tietenberg, Thomas H. (1996), *Environmental and Natural Resources Economics*, 4th edn, New York: HarperCollins.

Triche, Thelma A. (1990), 'Private Participation in the Delivery of Guinea's Water Supply Services', *Working Paper*, Infrastructure and Urban Development Department, Washington, DC: World Bank (August).

Trouvelot, Sandrine (1998), 'L'eau pour tous', *Alternatives Economiques*, 161 (July–August), 30–9.

United Nations (1973), *Multinational Corporations in World Development*, Department of Economic and Social Affairs, New York: United Nations.

United Nations (1973), *The Impact of Multinational Corporations on Development and on International Relations*, New York: United Nations.

United Nations Centre on Transnational Corporations (UNCTC) (1991), *World Investment Report 1991: The Triad in Foreign Direct Investment*, New York: United Nations.

United Nations Children's Fund (UNICEF) (1995), *UNICEF Strategies in Water and Environmental Sanitation*, New York: UNICEF (Document E/ICEF/1995/17, issued 13 April 1995).

United Nations Commission on Sustainable Development (UNCSD) (1997), *Comprehensive Assessment of the Freshwater Resources of the World, Report of the Secretary General*, Geneva, Commission on Sustainable Development, Fifth Session (7–25 April) (E/CN.17/1997/9) (February).

United Nations Commission on Sustainable Development (UNCSD) (2000), *Progress Made in Providing Safe Water Supply and Sanitation for all During the 1990s, Report of the Secretary-General*, Geneva, Eighth Session (24 April–5 May) (E/CN.17/2000/1) (14 March).

United Nations Committee on Natural Resources (UNCNR) (1994), *Review of the Progress on Water-Related Issues. Water Resources: Progress in the Implementation of the Mar del Plata Action Plan and of Agenda 21 on Water-Related Issues. Report of the Secretary General*, Geneva, Second Session (22 February–4 March) (E/C.7/1994/4) (12 January).

United Nations Conference on Trade and Development (UNCTAD) (1993), *World Investment Report 1993: Transnational Corporations and Integrated International Production*, New York and Geneva: United Nations.

United Nations Conference on Trade and Development (UNCTAD) (1995), *World Investment Report 1995: Transnational Corporations and Competitiveness*, New York and Geneva: United Nations.

United Nations Conference on Trade and Development (UNCTAD) (2000), *World Investment Report 2000: Cross-Border Mergers and Acquisitions and Development*, New York and Geneva: United Nations.

United Nations Economic and Social Council (UN ECOSOC) (1997), *Coordination of the Policies and Activities of the Specialised Agencies and other Bodies of the United Nations: Freshwater, including Clean and Safe Water Supply and Sanitation, Report of the Secretary General*, Geneva, UN ECOSOC, Substantive session of 1997 (30 June–25 July 1997) (E/1997/100) (9 June 1997).

United Nations Economic Commission for Europe (1981), *Long-term Perspectives for Water Use and Supply in the ECE Region: Report on Recent Experience and Prospects for Time Horizons 1990 and 2000* (prepared under the auspices of the ECE Committee on water problems), New York: United Nations.

United Nations Economic Commission for Latin America and the Caribbean (1990), *The Water Resources of Latin America and the Caribbean: Planning, Hazards and Pollution*, Santiago de Chile: United Nations.

UNDP/World Bank (1997a), *Water Vending in Old Dhaka, Balancing Inequities and Making Profits*, New Delhi, UNDP/World Bank South Asia.

UNDP/World Bank (1997b), *Workshop on the Willingness to Pay for Drinking Water Supply and Sanitation Report*, New Delhi: UNDP/World Bank South Asia (September).

UNDP/World Bank (1999a), *Improving Water Services Through Small Scale Private Providers, Water Vending in Chennai (Madras)*, New Delhi: UNDP/World Bank South Asia (June).

UNDP/World Bank (1999b), *Water and Sanitation Services for the Poor: Innovating Through Field Experience, Program Strategy 1999–2003*, Washington, DC: UNDP – World Bank Water and Sanitation Program.

UNDP/World Bank (1999c), *Water for India's Poor, Who Pays the Price for Broken Promises?*, New Delhi: UNDP/World Bank South Asia (October).

UNDP/World Bank (1999d), *Willing to Pay but Unwilling to Charge, do Willingness to Pay Studies Make a Difference?*, Field Note, New Delhi: UNDP/World Bank South Asia (June) 1999.

UNESCO (1987), *The Role of Water in Socio-economic Development: Report 1 of IHP-II Project C1 (on heightening awareness of the socio-economic role of water)/* prepared for the International Hydrological Programme by the Working Group of Project C1 (IHP-II), ed. W. E. Cox, Paris: UNESCO.

Union of International Associations (2000), *Yearbook of International Organizations 2000/2001: Guide to global civil society network*, 1B, Munich: K. G. Saur, Volume 1B.

Utting, Peter and Jaubert, Ronald (sous la direction de) (1998), *Discours et réalités des politiques participatives de gestion de l'environnement: le cas du Sénégal*, Geneva: IUED, UNSRID.

Vajpeyi, Dhirendra K. (ed.) (1998), *Water Resource Management, A Comparative Perspective*, Westport, Conn. and London: Praeger.

Valiron, François and Roche, Pierre-Allain (2000), 'L'Académie de l'Eau et la charte sociale', *L'eau au XXIème siècle: De la Vision à l'Action*, Paris: Editions Futuribles (March).

Van den Berg, Caroline (1997), 'Water Privatisation and Regulation in England and in Wales', *Public Policy for the Private Sector*, Note 115, Washington, DC: World Bank (May).

Vernon, Raymond (1992), 'Transnational Corporations: Where are they Coming from, Where are they Headed?', *Transnational Corporations*, 1(2), 7–35.

Vernon, Raymond (1996), 'International Investment and International Trade in the Product Cycle', *Quarterly Journal of Economics*, 80, 190–207.

Wahl, Richard W. (1993), 'Water Marketing in California: Past Experiences, Future Prospects', Los Angeles: Reason Foundation, *Policy Study*, 162 (July).

WaterAid (1999), *How Can WaterAid Respond to the Context in Which it Works? Private Sector Participation in the Water and Sanitation Sector: Challenges for WaterAid*, London: WaterAid.

Webb, Michael and Ehrhardt, David (1998), 'Improving Water Services Through Competition', *Public Policy for the Private Sector*, 164, Washington, DC: World Bank (December).

Wheat, Andrew (1996), 'Mexico Privatization, Pinata', *Multinational Monitor*, 17 (October).

Winpenny, James (1994), *Managing Water as an Economic Resource*, London and New York: Routledge.

World Bank (1981), *Annual Report*, Washington, DC: World Bank.

World Bank (1992a), *World Development Report 1992: Development and the Environment*, New York: Oxford University Press.

World Bank (1992b), *Water Supply and Sanitation Projects, the Bank's Experience: 1967–1989*. Operation Evaluation Department. Sector Study 10789. 06/19/92, Washington, DC: World Bank.

World Bank (1993), *Water Resources Management*. A World Bank Policy Paper, International Bank for Reconstruction and Development, Washington, DC: World Bank.

World Bank (1994a), *A Strategy for Managing Water in the Middle East and North Africa*, Washington, DC: World Bank.

World Bank (1994b), *Implementing the Water Resource Mandate of Agenda 21: the Promise and the Challenges for OECD Countries*, Transportation, Water, and Urban Development Department, Washington, DC: World Bank (April).

World Bank (1994c), 'Managing Urban Water Supply and Sanitation: Operation and Maintenance', *Lessons and Practices*, 5, Operations Evaluation Department, Washington, DC: World Bank (November).

World Bank (1994d), *World Development Report 1994: Investing in Infrastructure*, New York: Oxford University Press.

World Bank (1995a), *Kingdom of Morocco, Water Sector Review*, Washington, DC: World Bank, Report 14750–MOR (June).

World Bank (1995b), *The German Water and Sewerage Sector*, Transportation, Water, and Urban Development Programme, Washington, DC: World Bank (February).

World Bank (1997a), *The Private Sector in Infrastructure: Strategy, Regulation and Risk*, Washington, DC: World Bank.

World Bank (1997b), *Toolkits for Private Participation in Water and Sanitation*, Washington, DC: World Bank.

World Bank (1997c), *World Development Report 1997: The State in a Changing World*, New York: Oxford University Press.

World Bank (1997d), *Yemen, Towards a Water Strategy: An Agenda for Action*, Washington, DC: World Bank, Report 15718–YEM (13 August).

World Bank (1999a), *The Private Sector in Water, Competition and Regulation*, Finance, Private Sector, and Infrastructure Network, Washington, DC: World Bank.

World Bank (1999b), *Environment Matters at the WB*, Annual Review, Toward Environmentally and Socially Sustainable Development, Washington, DC: World Bank.

World Bank (1999c), *Annual Report*, Washington, DC: World Bank.

World Bank (2000a), *World Development Report 2000: Attacking Poverty*, Oxford: Oxford University Press.

World Bank (2000b), *Partnerships for Development*, Washington, DC: World Bank, 167–8.

World Bank Bangladesh (1998), *Water Resources Management in Bangladesh: Steps Towards a New National Water Plan*. Rural Report 17663, BD, Rural Development Sector Unit, South Asia Region, Dhaka: World Bank Dhaka Office.

World Commission on Dams (WCD) (2000), *Dams and Development: A New Framework for Decision-Making*, London: Earthscan (November).

World Water Council (1996), reprinted from *Hydropower and Dams*, Issue Four.

World Water Council (1997), *World Water Council Newsletter*, 0, Marseilles: WCC (March).

Wright, Vincent (sous la dir. de) (1993), *Les privatisations en Europe: programmes et problèmes*, Arles: Actes Sud; Poitiers: Observatoire du changement social en Europe occidentale.

Yakubu, N. (1997), *Water Supply Services: Kaduan State, Nigeria*, Institutional Development Series, The Water, Engineering and Development Centre (WEDC), Loughborough University.

Yepes, Guillermo (1999), *Do Cross-Subsidies Help the Poor to Benefit from Water and Wastewater Services? Lessons from Guayaquil*, Washington, DC: UNDP/World Bank, Water and Sanitation Program (February).

Yoskowitz, David W. (1999), 'Spot Markers for Water along the Texas Rio Grande: Opportunities for Water Management', *Natural Resources Journal*, 39(2).

Young, Oran and Hunold, C. (1999), 'Environment–Global Governance: Drawing Insights from the Environmental Experience', *Journal of Policy Analysis and Management*, 18(2).

Young, Robert Alton (1996), *Measuring Economic Benefits for Water Investments and Policies*, Washington, DC: World Bank.

Young, Z. and Boehmer-Christiansen, S. (1999), *The Functioning of the Global Environment Facility: A Political Analysis*, Economic and Social Research Council Final Report, United Kingdom (July).

Zajc, Katarina (1996), *Private Sector Participation Options in the Water Sector in Transition Economies*, George Mason University, UMI Publications.

Index